# Beyond Productivity

## Information Technology, Innovation, and Creativity

Committee on Information Technology and Creativity

Computer Science and Telecommunications Board
Division on Engineering and Physical Sciences

NATIONAL RESEARCH COUNCIL
OF THE NATIONAL ACADEMIES

William J. Mitchell, Alan S. Inouye, and Marjory S. Blumenthal, *Editors*

THE NATIONAL ACADEMIES PRESS
Washington, D.C.
**www.nap.edu**

THE NATIONAL ACADEMIES PRESS
500 Fifth Street, N.W., Washington, DC 20001

NOTICE: The project that is the subject of this report was approved by the Governing Board of the National Research Council, whose members are drawn from the councils of the National Academy of Sciences, the National Academy of Engineering, and the Institute of Medicine. The members of the committee responsible for the report were chosen for their special competences and with regard for appropriate balance.

Support for this project was provided by the Rockefeller Foundation. Any opinions, findings, conclusions, or recommendations expressed in this material are those of the authors and do not necessarily reflect the views of the sponsor.

International Standard Book Number 0-309-08868-2
Library of Congress Control Number 2003103683

Cover design by Jennifer M. Bishop

Copies of this report are available from the National Academies Press, 500 Fifth Street, N.W., Lockbox 285, Washington, DC 20055, (800) 624-6242 or (202) 334-3313 in the Washington metropolitan area. Internet, http://www.nap.edu.

# THE NATIONAL ACADEMIES
*Advisers to the Nation on Science, Engineering, and Medicine*

The **National Academy of Sciences** is a private, nonprofit, self-perpetuating society of distinguished scholars engaged in scientific and engineering research, dedicated to the furtherance of science and technology and to their use for the general welfare. Upon the authority of the charter granted to it by the Congress in 1863, the Academy has a mandate that requires it to advise the federal government on scientific and technical matters. Dr. Bruce M. Alberts is president of the National Academy of Sciences.

The **National Academy of Engineering** was established in 1964, under the charter of the National Academy of Sciences, as a parallel organization of outstanding engineers. It is autonomous in its administration and in the selection of its members, sharing with the National Academy of Sciences the responsibility for advising the federal government. The National Academy of Engineering also sponsors engineering programs aimed at meeting national needs, encourages education and research, and recognizes the superior achievements of engineers. Dr. Wm. A. Wulf is president of the National Academy of Engineering.

The **Institute of Medicine** was established in 1970 by the National Academy of Sciences to secure the services of eminent members of appropriate professions in the examination of policy matters pertaining to the health of the public. The Institute acts under the responsibility given to the National Academy of Sciences by its congressional charter to be an adviser to the federal government and, upon its own initiative, to identify issues of medical care, research, and education. Dr. Harvey V. Fineberg is president of the Institute of Medicine.

The **National Research Council** was organized by the National Academy of Sciences in 1916 to associate the broad community of science and technology with the Academy's purposes of furthering knowledge and advising the federal government. Functioning in accordance with general policies determined by the Academy, the Council has become the principal operating agency of both the National Academy of Sciences and the National Academy of Engineering in providing services to the government, the public, and the scientific and engineering communities. The Council is administered jointly by both Academies and the Institute of Medicine. Dr. Bruce M. Alberts and Dr. Wm. A. Wulf are chair and vice chair, respectively, of the National Research Council.

**www.national-academies.org**

# COMMITTEE ON INFORMATION TECHNOLOGY AND CREATIVITY

WILLIAM J. MITCHELL, Massachusetts Institute of Technology, *Chair*
STEVEN ABRAMS, IBM T.J. Watson Research Center
MICHAEL CENTURY, Rensselaer Polytechnic Institute
JAMES P. CRUTCHFIELD, Santa Fe Institute
CHRISTOPHER CSIKSZENTMIHALYI, MIT Media Laboratory
ROGER DANNENBERG, Carnegie Mellon University
TONI DOVE, Independent Artist, New York City
N. KATHERINE HAYLES, University of California at Los Angeles
J.C. HERZ, Joystick Nation Inc.
NATALIE JEREMIJENKO, Yale University
JOHN MAEDA, MIT Media Laboratory
DAVID SALESIN, University of Washington; Microsoft Research
LILLIAN F. SCHWARTZ, Computer Artist-Inventor, Watchung, New Jersey
PHOEBE SENGERS, Cornell University
BARBARA STAFFORD, University of Chicago

## *Staff*

ALAN S. INOUYE, Study Director and Senior Program Officer
MARJORY S. BLUMENTHAL, Director, Computer Science and
    Telecommunications Board
DAVID PADGHAM, Research Associate
MARGARET MARSH HUYNH, Senior Project Assistant
LAURA OST, Consultant
DAVID WALCZYK, Consultant
SUSAN MAURIZI, Senior Editor
JENNIFER M. BISHOP, Senior Project Assistant

# Preface

Computer science has drawn from and contributed to many disciplines and practices since it emerged as a field in the middle of the 20th century. Those interactions, in turn, have contributed to the evolution of information technology: New forms of computing and communications, and new applications, continue to develop from the creative interaction of computer science and other fields. Focused initially on interactions between computer science and other forms of science and engineering, the Computer Science and Telecommunications Board (CSTB) began in the mid-1990s to examine opportunities at the intersection of computing and the humanities and the arts. In 1997, it organized a workshop that illuminated the potential, as well as the practical challenges, of mining those opportunities[1] and that led, eventually, to the project described in this report. Ensuing discussions between CSTB staff and people interested in the intersection of computing and the humanities or the arts, notably Joan Shigekawa of the Rockefeller Foundation, a participant in the 1997 workshop, culminated in a grant from the Rockefeller Foundation to study information technology and creativity (see Box P.1 for the statement of task).

This report should be read with two conditions in mind: First, it is, by design, a record of the project, filled with descriptions, observations, conclusions, and recommendations intended to motivate and sustain interest and activity in the rich intersection of information technology (IT) and the arts and design. Second, in this book form it cannot possibly convey the exciting possibilities at that intersection. Instead, it presents examples and pointers to sites on the World Wide Web and in the physical world where that intersection can be observed and experienced. We urge the reader to treat this report as a

---

[1]See *Computing and the Humanities: Summary of a Roundtable Meeting,* published in 1998 by the American Council of Learned Societies, one of three collaborators with CSTB in organizing the workshop.

---

**BOX P.1**

**Statement of Task**

A series of discussions among a cross section of the arts community and experts in computing and communications will be organized. These discussions will crystallize new ways of conceptualizing joint opportunities and new approaches to the arts (and/or IT [information technology]). They will explore what would make the most conducive environment for IT-arts exchange on an ongoing basis, considering physical and virtual options. They will address possible mechanisms to sustain the discussion, such as funding and institutional support. Finally, they will culminate in both a coherent description of potential futures and an agenda for action, action that bridges the different communities as well as action most appropriate for one or another.

---

primer and guidebook and to seek out instances of IT and creative practices—ITCP—directly.

## COMMITTEE COMPOSITION AND PROCESS

The study committee convened by CSTB featured an unusually eclectic group of individuals (see Appendix A for biographies of committee members). Characterizing most (or all) of them as experts on particular subjects would only begin to suggest the talents of this group. Collectively, the committee had expertise and experience in the intersections of information technology and music, the visual arts, film, and literature and in art history, architecture, cultural studies, and many of the technologies pertinent to ITCP. The committee did its work through its own deliberations and by soliciting input from a number of other experts (see Appendix B for a list of those who briefed the committee). It met first in August 2000 and five times subsequently in plenary session. Additional information was derived from reviewing the published literature, monitoring selected listservs and Web sites, and obtaining informal input at various conferences and other convenings. During the editorial phase of the study, facts were checked for accuracy with either authoritative published sources or subject experts.

The diversity of this committee made it a microcosm of some of the communities it hopes to influence with this report. That diversity posed challenges in the conduct of this project that will be echoed in attempts to learn from it: Conversations among people with different training and professional experience can be confounded by jargon and prejudices as well as by differing knowledge bases—even when those people share interests. The completion of this report attests to the potential for technologists and artists to find common ground, not only in undertaking creative work, but also in contemplating options for making such work easier to undertake and more widespread. But finding this common ground sometimes proved to be a formidable challenge.

The productive interaction among committee members was captured in some of their career developments during the course of this project. Chris Csikszentmihalyi, for example, left Rensselaer Polytechnic Institute to join John Maeda at MIT's Media Lab. Michael Century left McGill University for Rensselaer Polytechnic Institute. Natalie Jeremijenko was hosted by Jim Crutchfield for a month's professional visit at the Santa Fe Institute. And John Maeda was inspired by the project to build "a new online Bauhaus." These and other developments attest to the dynamism and creative energy of the people who have been exploring the intersection of IT and creativity.

Although the report refers to several companies, products, and services by name, such reference does not constitute an endorsement by the committee or the National Academies. The committee did not evaluate any product or service in sufficient detail to allow such an endorsement.

• • • • • • • • • • • • • • • • • • • • • • • • • • • • • • • • • • • •

# ACKNOWLEDGMENTS

The committee is particularly grateful to Joan Shigekawa of the Rockefeller Foundation for initiating this study. She approached CSTB with a conviction that the time was right for a conversation among people of different backgrounds about how to enhance and sustain the intersection of information technology and creative practices. We appreciate her guidance and support through the study process, including her participation in two committee meetings, occasional relay of useful information, and continuing demonstration of interest in the process and the eventual results.

In addition, we would like to thank those individuals who provided valuable inputs into the committee's deliberations. Those who briefed the committee at one of our plenary meetings are listed in Appendix B. Others who provided us with important inputs include Bill Alschuler (California Institute of the Arts), Howard Besser (New York University), Shari Garmise (Consultant, Washington, D.C.), Samuel Hope (National Office for Arts Accreditation), Sharon Kangas (Center for Arts and Culture), Anna Karlin (University of Washington), Ruth Kovacs (The Foundation Center), Joan Lippincott (Coalition for Networked Information), and Laurens R. Schwartz (Consultant, New York City). We would also like to acknowledge those organizations that hosted committee meetings: the American Institute of

Graphic Arts, New York University, Stanford University, Pixar Animation Studios, and the Massachusetts Institute of Technology.

The committee appreciates the thoughtful comments received from the reviewers of this report and the efforts of the National Research Council's report review coordinator. The review draft stimulated a comparatively large volume of comments, many of which provided additional reference material, relevant anecdotes, and observations to bolster or counter the committee's earlier thinking. The comments were instrumental in helping the committee to sharpen and improve this report. In particular, Simon Penny of the University of California at Irvine provided an unusually extensive and thoughtful set of comments that served to improve the quality of this final report.

Finally, the committee would like to acknowledge the staff of the NRC for their work. Alan Inouye served as the study director with overall staff responsibility for the conduct of the study and the development of this final report; his effort to bring the report to completion was exceptional and demanded far more of his time than anticipated. Marjory Blumenthal, director of the CSTB, provided essential guidance and input throughout the study process, drafted and edited a number of sections of the final report, and was both helpful and patient in bringing the committee process to a successful conclusion. Margaret Marsh Huynh had primary responsibility for the administrative aspects of the project such as organizing meeting logistics; her efforts made a particularly complicated and demanding process run smoothly. Consultants Laura Ost and David Walczyk generated initial drafts of several sections of the report; Ms. Ost also edited several chapters. Susan Maurizi edited the manuscript for publication. David Padgham and Jennifer Bishop provided research assistance; Ms. Bishop also created several of the original figures that appear in this report (including the cover design). The committee also thanks Janet Briscoe, Janice Sabuda, and Brandye Williams of the CSTB, and Claudette K. Baylor-Fleming and Carmela J. Chamberlain of the Space Studies Board for their support of the committee's work.

William J. Mitchell, *Chair*
Committee on Information Technology and Creativity

# Acknowledgment of Reviewers

This report has been reviewed in draft form by individuals chosen for their diverse perspectives and technical expertise, in accordance with procedures approved by the National Research Council's Report Review Committee. The purpose of this independent review is to provide candid and critical comments that will assist the institution in making its published report as sound as possible and to ensure that the report meets institutional standards for objectivity, evidence, and responsiveness to the study charge. The review comments and draft manuscript remain confidential to protect the integrity of the deliberative process. We wish to thank the following individuals for their review of this report:

Anna Bentkowska, Conway Library, Courtauld Institute of Art,
Howard Besser, New York University,
Sandra Braman, University of Alabama,
Donna Cox, University of Illinois at Urbana-Champaign,
Robert Denison, First Security Company,
Steve Dietz, Walker Art Center,
Kristian Halvorsen, Hewlett Packard Laboratories,
Paul Kaiser, Independent Artist, New York City,
Alan Kay, Hewlett Packard Company,
Clifford Lynch, Coalition for Networked Information,
Simon Penny, University of California at Irvine,
Bill Seaman, Rhode Island School of Design, and
Mark Tribe, Rhizome.org.

Although the reviewers listed above have provided many constructive comments and suggestions, they were not asked to endorse the conclusions or recommendations, nor did they see the final draft of the report before its release. The review of this report was overseen by Edward Lazowska, University of Washington. Appointed by the

National Research Council, he was responsible for making certain that an independent examination of this report was carried out in accordance with institutional procedures and that all review comments were carefully considered. Responsibility for the final content of this report rests entirely with the authoring committee and the institution.

# Contents

# Summary and Recommendations

Creativity plays a crucial role in culture; creative activities provide personal, social, and educational benefit; and creative inventions ("better recipes, not just more cooking") are increasingly recognized as key drivers of economic development. But creativity takes different forms at different times and in different places. This report argues that, at the beginning of the 21st century, information technology (IT) is forming a powerful alliance with creative practices in the arts and design to establish the exciting new domain of information technology and creative practices—ITCP. There are major benefits to be gained from encouraging, supporting, and strategically investing in this domain.

• • • • • • • • • • • • • • • • • • • • • • • • • • • • • • • • • • • • • • • • • •

## INFORMATION TECHNOLOGY AND CREATIVE PRACTICES

Alliances of technology and creative practices have often emerged in the past. In the 19th century, for example, optical, chemical, and thin-film manufacturing technologies converged with the practices of the pictorial arts to establish the new domain of photography. Then, photographic technology became further allied with the practices of the performing arts, giving rise to the domain of film. The cultural and economic consequences of these developments have been profound. The emerging alliance of information technology with the arts and design has, this committee believes, even greater potential.

ITCP has already yielded results of astonishing variety and significant cultural and economic value. These results have taken such forms as innovative architectural and product designs, computer animated films, computer music, computer games, Web-based texts, and interactive art installations, to name just a few. They have developed from individual, group, and institutional activities; the processes by which they have been produced have spanned both the commercial and not-for-profit worlds and the formal and informal economic sectors. The products of ITCP have begun to appear in many different countries, in ways that reflect cultural, economic, and political differences.

IT has now reached a stage of maturity, cost-effectiveness, and diffusion that enables its effective engagement with many areas of the arts and design—not just to enhance productivity or to allow more efficient distribution, but to open up new creative possibilities. There is a highly competitive race for leadership in this domain. The potential payoffs from success in the near- and long-term futures are enormous: billion-dollar industries, valuable exports, thriving communities that attract the best and the brightest, enriched cultural experiences for individuals and communities, and opportunities for global cultural visibility and influence.

By definition, there is no formula for creativity. But there are effective ways to invest in establishing conditions necessary for ITCP, in overcoming impediments, and in providing incentives. Furthermore, there are ways to recognize and reward creative contributions and to derive social benefit from them. In appropriate combination, these measures can add up to powerful strategies for encouraging, supporting, and reaping the rewards of ITCP. Development along with implementation of such strategies is the challenge addressed by this report.

## MULTILEVEL STRATEGIES FOR ITCP

ITCP can be engaged at multiple levels—by individual artists and designers who deal with IT tools, media, and themes; in the structuring and management of cross-disciplinary research and production groups working in the ITCP domain; in directing educational and cultural institutions with interests in ITCP; at the level of regional development strategy aimed at fostering ITCP clusters; as an aspect of national economic and cultural policy; and in multinational collaborative efforts. All of these levels are important, and there are cross-connections among them. There is, therefore, considerable advantage in coordinated, multilevel strategies for encouraging, supporting, and benefiting from ITCP.

# PROVIDING NEW TOOLS AND MEDIA FOR ARTISTS AND DESIGNERS

Individual artists and designers have experimented with IT since its earliest incarnations. Artistic exploration of the possibilities of computer graphics, for example, now extends back more than 30 years, and 40 years for computer music. As IT has matured and been assimilated into the mass market, the IT tools and media available to artists and designers have become both more diversified and more affordable. There are popular, standardized tools for performing such tasks as creating, editing, and distributing images, audio, and text; there are variants on standard tools customized to the needs of particular artists or designers; and there are highly specialized, purpose-built tools used by nobody but their creators.

To a software developer or an information services manager, it might seem that the keys to ITCP are simply equipment and software—developing and providing access to standard, commercial IT tools for artists and designers. This perspective is useful as far as it goes, and it can provide a good way to get started with ITCP, but in the long run it is an insufficiently rich or flexible one. We make our tools; then our tools make us.[1] Furthermore, software tools encode numerous assumptions about the making of art and design—precisely the sorts of presuppositions that truly creative practitioners will want to challenge. And the more software tools emphasize ease of use or familiar metaphors, the more they must depend on restrictive assumptions in order to do so. Such tools not only must be available, but they also must be objects of critical reflection; they must be open to adjustment and tweaking, they must support unintended and subversive uses—not just anticipated ones—and they must not be too resistant to being torn apart and reconceived. If creative practice can develop the powerful spaces and tools that it needs, like the electronic easel or electronic studio, these spaces and tools could help transform or enlarge the metaphors, spaces, and tools (office, desktop, files) that the rest of us have to work with.

The relationship between IT professionals and artists and designers will be of limited value if it is conceived simply as one of software (or hardware) producer and consumer. It should, instead, be one of flexible and thoughtful collaboration in which the roles of software designer and user are not rigidly distinguished. The advances made by IT researchers may suggest new forms of art and design practice,

---

[1]Inspired by Marshall McLuhan, 1954, "Notes on the Media as Art Forms," *Explorations* 2 (April): 6-13.

while the questions raised by artists and designers may provide new ways of thinking about IT—ITCP work challenges the boundaries of traditional disciplines. Modular, reusable and recombinable code elements may support critical reconceptualization more readily than closed, proprietary software products. Open source development may provide better opportunities for cross-disciplinary collaboration, customization, and reconceptualization than tools developed and marketed as protected intellectual property—no matter how powerful and attractive those tools may be.

## PROVIDING OPPORTUNITIES TO DEVELOP ITCP SKILLS

In general, ITCP depends on opportunities for learning across multiple disciplines—some mix of the arts and design plus IT concepts and tools. The growing numbers of artists and designers becoming skilled programmers or hardware developers, like the smaller number of computer scientists and technologists engaging seriously with the arts and design, demonstrates that this is feasible. But it is not easy: Colleges and universities focus mostly on established disciplines, and the cross-disciplinary programs that do exist vary widely in their institutional support, effectiveness, and quality.

Like other professionals, artists and designers can do more with IT if they become deeply conversant with its capabilities and limitations. Achieving that result requires far more than training on standard tools, and it also demands an ability to understand tools and media critically—in cultural and historical context. Such critical thinking about tools is much less typical of education and training in IT, a difference that contributes to the asymmetric participation of artists and computer scientists in ITCP. To date, it seems that artists and designers have made greater efforts to engage IT seriously than computer scientists and technologists have made to acquire deep understanding of creative practices in the arts and design. It is easier to find designers who can program than programmers who can design, or composers comfortable with signal processing than specialists in signal processing who can compose or perform at high levels of proficiency. This imbalance could change, with outreach to the computer science community and interest in ITCP among those who provide funding and other incentives and rewards.

Although motivated individuals can and do acquire complementary IT and arts or design skills, significant ITCP work can also be produced by cross-disciplinary partnerships between computer scientists and artists or designers. This approach has the advantage of requiring that fewer skills be mastered by individual team members, and it is often essential for large projects, but there are some inherent difficulties. Progress in collaborative ITCP requires effective dialogue between artists and designers and IT professionals. Differences in professional culture, styles, and values, as well as communication problems, can confound effective collaboration. Yet there are strong traditions of successful cross-disciplinary collaboration in architecture (particularly as computer-aided design/computer-aided manufacturing (CAD/CAM) technology plays an increasing role), in film production, and in the creation of video games, and there have been some successful pairings of artists and technologists to produce visual works, performances, and installations.

## CREATING ENVIRONMENTS THAT SUPPORT ITCP

ITCP work can be done in many different places. And the diversity of venues matters, since each type of venue represents different tradeoffs and provides different combinations of opportunities, constraints, and comparative advantage. So an effective ITCP development strategy is likely to be a multivenue one.

ITCP venues may occupy physical or virtual spaces, be large or small, range from loosely organized collectives to formal programs, and be either free-standing or connected to established institutions. Specialized exhibitions, performance festivals, presentation and lecture series, conferences, Internet forums, and display and performance sites have all played important roles in the growth of ITCP communities. By contrast, mainstream arts and design organizations—museums, galleries, arts and design fairs, arts and design publishers, and so on—have played a lesser role, although they have begun to embrace ITCP more as the products of ITCP have played a larger cultural role and as these products have developed in quality and interest.

Much pioneering exploration of ITCP has taken place in studio-laboratories, which build on the tradition of earlier centers of cross-disciplinary research and education in the arts, design, and new technology of the time, such as Germany's Bauhaus in the pre-World War II years, the postwar New Bauhaus in Chicago, and the Center for Advanced Visual Studies established by Gyorgy Kepes at the Massachusetts Institute of Technology (MIT) in the 1960s. MIT's Media Laboratory has been among the largest and most visible, and it has generated affiliates in Europe and Asia. However, the Media Lab's combination of substantial laboratory and human resources with an atelier style of research and education, building on a consortium of industry funders, is difficult to replicate outside the context of a leading research university with strong industrial connections. Some universities, such as Carnegie Mellon University, have formed special cross-disciplinary centers that undertake ITCP, and several arts schools, such as the California Institute of the Arts and the Art Center College of Design in Pasadena, have transformed their curricula to incorporate

IT, yielding numerous focused ITCP activities. Some film schools have shifted their emphasis from traditional to digital production and distribution technologies, and most architecture and design schools have supplemented or supplanted drawing boards with CAD. Several universities have begun to develop cross-disciplinary study programs in aspects of ITCP. But a key challenge, particularly in times of tight finances, is to find effective ways to fund these programs—and to frame them in ways that are pedagogically sound and appropriately adaptive to the continuing evolution of ITCP.

In Canada and Europe, and emerging in Asia and Australia, major efforts are under way to develop standalone, government-backed ITCP centers. Such centers are typically conceived of as instruments of arts and cultural policy, rather than as equivalents of national research laboratories. This is an arena in which the United States lags. In principle, such centers can provide considerable flexibility and freedom of intellectual direction. On the down side, they are vulnerable to changes in government spending priorities, they can lose the very independence that makes them attractive if they shift to executing contracts from industry, and they are usually less able to draw effectively on the laboratories and human resources of large universities.

The technology required for ITCP can be expensive, and ambitious ITCP productions can require major funding. Given the breadth of ITCP, some funding is available through commercial channels. It normally requires close engagement with popular culture and mass audiences, with all the constraints and opportunities that this implies. This path is illustrated by the film and entertainment industries—these ITCP pioneers overcame difficulty and expense and now can produce major commercial successes. A focused example is the flourishing video game industry, a direct outcome of the rise of ITCP. It obviously would not be possible at all without the necessary IT, and its products define a new art form that also resonates with the general public. It has found some highly innovative ways to combine centralized research, development, and marketing with large-scale open-source strategies, and it has evolved unique distribution strategies.

Operating on a small scale and often producing innovative work through commissions from enlightened patrons is another group of players that straddle the boundary between commerce and the arts: Independent architectural design, product design, graphic design, and music and video production houses now make extensive use of IT tools and media, and they frequently have IT specialists on staff. In some cases, this amounts to little more than straightforward use of standard, commercial tools. But more adventurous and innovative houses have seized the opportunity, through IT, to open up some exciting new domains. This is particularly evident in the move of architects into CAD/CAM design and construction—with the resulting emergence of new architectural idioms—and the move of graphic designers into work that is more interactive.

Much important ITCP work occurs outside the marketplace. In addition to academic efforts, individual, independent artists and designers, operating mostly on a small scale, are responsible for a crucial

segment of ITCP. By virtue of their independence, they are well positioned to provide perspectives that challenge mainstream thinking
and to engage industry as catalytic outsiders who can instigate new
ways of thinking about products and processes. Many forms of traditional art production, such as painting and writing, are labor-intensive
and modest in their requirements for investments in technology, but
ITCP is often much more capital-intensive. This increased need for
capital presents a chronic problem for independents; they often operate on a shoestring, struggle to get access to technology and expertise,
and must make whatever technology investments they can manage
from project-by-project funding. They usually depend on some mix of
the gallery and patronage structures of the art world, arts foundation
grants, and relationships with sympathetic educational institutions
and corporations.

ITCP activity in all of these venues tends to cluster geographically.
Fostering such clusters—with a vital mix of commercial, non-profit,
academic, design and production house, and independent practitioner
activity—can play an important role in regional economic development. There can be major direct benefits to local economies, and indirect (but potentially even more important) benefits in the form of
better design and higher levels of innovation distributed over many
sectors of the economy.

In addition, by its very nature, ITCP lends itself to efficient electronic connection of scattered islands of activity. Writers and photographers can submit their work electronically to distant publishers,
architects can form geographically distributed design and construction teams, film studios in Hollywood can link electronically to
postproduction houses in London or animation shops in Korea, and so
on. That capability for connectivity is leading, increasingly, to multinational ITCP alliances and organizations. Such a capability can be
particularly important in contexts—such as in developing nations—
where the local culture supports some unique ITCP cluster and electronic connectivity adds value to that cluster by providing wider
access to resources and markets. It is also important in contexts—such
as those of Australia, New Zealand, and Singapore—where small but
highly educated populations, combined with the effects of distance,
make concentration on high-value, immaterial, information goods and
services particularly attractive.

• • • • • • • • • • • • • • • • • • • • • • • • • • • • • • • • • •

# FOSTERING THE CULTURE OF INFORMATION TECHNOLOGY AND CREATIVE PRACTICES

Providing new tools and media for artists and designers, providing opportunities to develop ITCP skills, and creating environments
that support ITCP are all necessary to form thriving ITCP clusters, but

they are not in themselves sufficient. It is also essential to foster the culture of ITCP—the flow and exchange of ideas among those engaged, the development of a sense of intellectual community, the representation of ideals and values, and the recognition and validation of outstanding work.

The academic environment, in particular, is central to the future of ITCP. That is where talent is cultivated, and that is where research and practice of various kinds can take place largely without market strictures. At present, a gulf exists between computer science and the arts and design. Although some computer scientists bridge that gulf—and contribute considerably to ITCP—that activity often happens outside their department. Although some arts departments have been skeptical of "new-media" programs, in general the arts and design on campus have welcomed ITCP more than have computer science departments. The lack of welcome from computer science departments reflects a lack of appreciation of ITCP's potential to contribute to the advance of computer science as a field, as well as concern about already tight curricula. At the same time, arts and design departments on campuses and arts schools have sought to internalize ITCP facilities and to develop their own research and teaching programs in ITCP. The situation echoes earlier efforts to formalize computer science as a field, establish a theoretical foundation for it, and provide it with some level of autonomy from its predecessor and sister fields. But it is important to explore the potential for constructive interaction between the arts and design and computer science before universities—and practitioners—conclude that "parallel play" is the way to go.

Building academic clusters is a nontrivial challenge. Not only are there cultural differences among the constituent disciplines, but there are also significant differences in expectations for funding, use of time, use of graduate students, definitions of what is acceptable work, and so on. Special centers, seminars, and other venues are being tried on campuses, a kind of institutional experimentation that is vital to developing ITCP. They help to frame and sustain ITCP projects. The time is ripe for academic experimentation with ITCP, from course content and curricula to institutional options and incentives.

Education, collaboration, funding, and professional advancement all depend on how ITCP is received. Because ITCP spans so many activities, there is feedback from the commercial space and popular culture—a powerful reinforcement on the design end—and there is more ambiguous feedback through academic institutions (faculty and administrators); publications, exhibitions, performances, and prizes, as well as those who select for them; and funders of research and the arts.

Because the field of ITCP is young and dynamic, ITCP production is hard to evaluate. Traditional review panels—representing funders; owners and managers of conventional display, performance, or publication outlets; and those making personnel decisions at academic institutions—may be hampered by their members' ties to single disciplines and the absence of a time-tested consensus about what constitutes good work in ITCP and why. This problem is typical of new fields drawing from multiple disciplines, albeit aggravated by the contrast between computer science and the arts and design. It is offset somewhat by a flourishing array of conferences and other forums, in both virtual and real space, that provide a sense of community and an outlet as well as feedback. Effective evaluation, validation, and recognition of ITCP work are essential for this domain to progress. Building on traditions in the arts and design, prizes can be powerful for stimulating and recognizing excellence in ITCP.

# A NEW FORM OF RESEARCH

ITCP can constitute an important domain of research. It is inherently exploratory and inherently transdisciplinary.[2] Concerned at its core with how people perceive, experience, and use information technology, ITCP has enormous potential for sparking reconceptualization and innovation in IT. In execution, it pushes on the boundaries of both IT and the arts and design. Computer science has always been stimulated by exposure to new points of view and new problems, which are ever-present in the arts and design. Because of the breadth of use to which artists and designers put different forms of IT, and because they typically are not steeped in conventional IT approaches, artists' and designers' perspectives on tools and applications may provide valuable insights into the needs of other kinds of IT users. The needs and wants of artists and designers can suggest new ways of designing and implementing IT. Engaging their perspectives is a logical extension of recent trends in cross-disciplinary computer science research.

Recently, for example, artists and designers have brought new concerns to the design and implementation of sensor systems, distributed control systems and actuators, generative processes and virtual reality, and the Internet and other networks. Their interests in performance and in engaging the public present challenges for system interactivity; their interests in improvisation present new opportunities for exploring human-machine interaction. Although artists and computer scientists have long interacted in such spheres as computer graphics and music, almost any form of IT may be adopted or adapted for uses in the arts and design. This flexibility of purpose parallels the plasticity of the computer itself—and that helps to explain why artists' concerns may motivate new combinations as well as new forms of IT.

It is important to recognize, however, that serious ITCP research goes beyond appropriation of established IT concepts and techniques for artistic or design purposes, or use of straightforward examples

---

[2]In transdisciplinary ITCP work, artists and designers interact as peers with computer scientists, a model that is described in detail in Chapter 4.

drawn from the arts and design to demonstrate the potential applications of new IT. It requires drawing on deep understanding of *both* IT and the arts and design to formulate scientifically interesting new questions in ITCP, and to see the subtle cultural implications of relevant new science. Issues arising from the arts and design have motivated challenging and important domains of computer science and technology research, such as three-dimensional geometric modeling and scene rendering directed at the practices and needs of designers and animators. Sometimes arts-oriented researchers raise cultural, social, ethical, and methodological questions for computer scientists that would not be obvious in a more narrowly focused technological context. Conversely, outcomes of computer science research may challenge artists and designers to rethink their established assumptions and practices (rethinking that includes an evolution from artifact creator to process mediator), as when architects engage the possibilities of curved-surface modeling and associated CAD/CAM fabrication techniques, or when photographers ponder the differences in the roles of digital and silver-based images as cultural products and as visual evidence. And there are areas, such as augmented reality, tangible computing, lifelike computer animation of characters, and user-centered evaluation of computer systems, that are probably best regarded as the joint outcomes of questions posed and investigations conducted by computer scientists and by artists and designers. These developments suggest that the value of ITCP lies not just in the capacity of each field to answer questions posed by the other, but also in the opportunity for each field to gain fresh, sometimes uncomfortable, perspectives on itself.

## MAKING ITCP HAPPEN

The broad scope of ITCP implies that it derives funding from both commercial activity—notably in design and entertainment contexts—and non-profit activity. The latter is where support is particularly uncertain yet essential, since it is in non-profit contexts that much experimentation takes place and some of the broadest public, participant access becomes possible. The hybrid nature of ITCP tends to confound its funding. In the United States, exploratory and productive work in the arts and at the non-commercial frontiers of design is likely to be funded by private philanthropy, while in computer science the leading funders of basic research are government agencies, often in support of specific agency missions. Computer science research grants are larger (by an order of magnitude) than grants (or prizes) typically available to artists—and they tend to be tied to the advances in scientific knowledge or the specific kinds of applications of concern to their funders.

Advancing ITCP requires new approaches to funding. A first step is recognition by both the arts and computer science patrons that topics in ITCP are legitimate; next must come support for exploration of the intersections between IT and the arts and design, and with that support for new kinds of technical and social and intellectual infrastructure for undertaking and providing access to ITCP. Those new approaches, in turn, may require new skills and participants in funders' decision-making processes. Grant program definitions should specifically embrace ITCP, but without that, progress in ITCP will depend on grant seekers' ingenuity in influencing program definitions and relating their ideas to existing programs.

In addition to monetary support, ITCP depends on resolving concerns about intellectual property rights. Not only does ITCP feature a broad range of content and a broad range of expression, but its production can also involve creative reuse or adaptation of previously generated content or expression. It also requires attention to the archiving and preservation of IT-based works, both those of a fixed nature and those designed to change through interactivity or other factors.

The rise of ITCP and the process of contemplating its future point to the need for better data on arts-related activities and trends. Although imperfect, the data available on scientific and technical research is better than that for arts activities. The lack of good data hinders effective planning and policy making.

• • • • • • • • • • • • • • • • • • • • • • • • • • • • • • • • • • • • • • •

# RECOMMENDATIONS

Realizing the potential of ITCP requires actions on many fronts—by individuals, organizations, and funders of different kinds. The benefits will accrue broadly—in multiple sectors of the economy, geographic regions, and disciplines. Other efforts already address the roles of established arts institutions—museums, galleries, theaters, and so on—in relation to IT-based art works and performances. This report concentrates its recommendations on those most responsible for nurturing the talent and the explorations that are the essence of ITCP. The recommendations below build on discussions in the body of the report, which explores the ecology of creative practices and the components of the strategies through which ITCP can thrive.

## FOR EDUCATORS AND ACADEMIC ADMINISTRATORS

1. Support the achievement of fluency in information technology (IT), and the development of critical and theoretical perspectives on IT, by arts and design students through the provision of suitable

facilities, opportunities for hands-on experience with IT tools and media, and curricula that engage critical and theoretical issues relating to IT and to information technology and creative practices (ITCP).

2. Support educational experiences for computer science students that provide direct experience in the arts and design, critical discussion, and formation of broader cultural perspectives—not merely as semi-recreational enrichment, but at a sufficiently challenging level to raise hard questions about the social and cultural roles both of science and technology and of the arts and design.

3. Foster exploration of ITCP through incentives and experimentation with a range of informal (e.g., workshops and seminars) and formal vehicles (e.g., centers, awards)—in particular, by building firmly and boldly on demonstrated local (and often small-scale) strengths and productive relationships already in place.

4. Support curricula, especially at the undergraduate level, that provide the necessary disciplinary foundation for later specialization in ITCP.

## For Foundations, Government Agencies, and Other Funders

5. Allocate funding not only to support work by specialists in established and recognized areas of IT and of the arts and design, but also to foster collaborations that open up new areas of ITCP.

6. Structure proposal review processes to encourage not only continued development of established and recognized areas of IT and of the arts and design, but also higher-risk, longer-horizon efforts to develop ITCP.

7. Provide program managers with more time and leeway to learn about new fields and new kinds of grantees; encourage mobility among grant makers, artists, designers, and computer scientists.

8. Develop a new grant-making category for tool (instrument) building, emphasizing designs that are extensible and tools that provide support for improvisation, and for providing broad access to the resulting tools. Expand research program support for work in aspects of distributed control, sensors and actuators, video and audio processing, human-computer interaction, information retrieval, artificial intelligence, networking, embedded systems, generative processes, and other technological areas that are critical to advancing ITCP, with a particular focus on arts-and-design-inspired applications of these technologies that extend beyond conventional uses.

9. Factor infrastructure and archiving and preservation needs into grant levels because this support is essential to enable future work in ITCP.

10. Support the establishment of new prizes for excellence in ITCP and the development of curated Web sites for its display or performance.

11. To support policy decision making, underwrite a better knowledge base—ranging from the history of ITCP to the details of who is doing what, where, when, and how—that parallels the knowledge base in scientific and engineering fields.

12. Underwrite research on the formation of creative clusters and the role that ITCP can play in promoting regional development.

13. Provide support for the creation and maintenance of networks of organizations (composed of participants from academia, industry, and cultural institutions) involved with ITCP.

## FOR INDUSTRY

14. Seek opportunities to develop new products and services relating to the growing field of ITCP and to participate in the formation of ITCP clusters.

15. Pursue relationships with centers of ITCP activity, and seek opportunities to engage artists and designers who can contribute to the development of ITCP products and services.

## FOR THE NATIONAL ACADEMIES

16. Organize a symposium series on Frontiers of Creative Practice (paralleling the Frontiers of Science and Frontiers of Engineering series) to bring together a cross section of young artists, designers, scientists, and technologists working within ITCP.

# 1 Information Technology, Productivity, and Creativity

The benefits of information technology (IT) extend far beyond productivity as it is usually understood and measured. Not only can the application of IT provide better ratios of value created to effort expended in established processes for producing goods and delivering services, but it can also reframe and redirect the expenditure of human effort, generating unanticipated payoffs of exceptionally high value. Information technology can support inventive and creative practices in the arts, design, science, engineering, education, and business, and it can enable entirely new types of creative production. The scope of IT-enabled creative practices is suggested (but by no means exhausted) by a host of coinages that have recently entered common language—computer graphics, computer-aided design, computer music, computer games, digital photography, digital video, digital media, new media, hypertext, virtual environments, interaction design, and electronic publishing, to name just a few.

The benefits of such practices have economic, social, political, and cultural components. IT-enabled creative practices have the potential to extend benefits broadly, not only to economic and cultural elites (where they are most immediately obvious), but also to the disadvantaged, and not only to the developed world but also to developing countries. And their impacts extend in two directions: Just as the engagement of IT helps shape the development of inventive and creative practices, so also can inventive and creative practices positively influence the development of IT. See Box 1.1. This report explores the complex, evolving intersections of IT with some important domains of creative practice—particularly in the arts and design—and recommends strategies for most effectively achieving the benefits of those intersections.

---

**BOX 1.1**

**The Utility of Information Technology**

A common answer to the question, What good is information technology?, is that it enhances productivity. Unquestionably, information technology (IT) now helps one to perform many routine tasks with greater speed and accuracy, with fewer errors, and at lower cost. So computers and software products are marketed as productivity tools, investments in IT are justified in terms of productivity gains, and economists try (sometimes without success) to measure those gains. In this role, IT is a servant.

An additional claim, which can be justified in certain contexts, is that IT enhances the quality of results. Laser-printed documents not only are quicker and cheaper to produce than handwritten or typewritten ones but may also be crisper and more legible. The outputs of detailed computer simulations of systems may be more reliable, and more useful to engineers, than the approximate, rule-of-thumb hand calculations that were used in earlier eras. And a sophisticated optimization program may produce a better solution to an allocation problem than manual trial and error. In this role, IT supports creative craftsmanship.

A still stronger, but frequently defensible, claim is that IT enables innovation—the production of outcomes that would otherwise simply not be possible. Scientists may use computers to analyze vast quantities of data and thereby derive new knowledge that would not be accessible by other means. Architects may use curved-surface modeling and computer-aided design/manufacturing systems to design and build forms that would have been infeasible—and probably would not even have been imagined—in earlier times. And new, electronic musical instruments, which make use of advanced sensor and signal-processing technology, may open up domains of composition and performance that could not be explored using traditional instruments. In this role, IT becomes a partner in processes of innovation.

Perhaps the strongest claim is that IT can foster practices that are creative in the most rigorous sense—scholarly, scientific, technological, design, and artistic practices that produce valuable results in ways that might be explained in retrospect but could not have been predicted. At this point, one might detect a whiff of paradox—a variant on Plato's famous *Meno* paradox. Unless it offers users a means to produce something they already know they want, IT is not helpful. But if someone produces something merely by running a program, the production process is predetermined and potentially standardized, so how can the result be truly creative?

---

# INVENTIVE AND CREATIVE PRACTICES

Creativity is a bit like pornography; it is hard to define, but we think we know it when we see it.[1] The complexities and subtleties of precise definition should not detain us here, but it is worth making a few crucial distinctions.

Certainly, creative intellectual production can be distinguished from the performance of routine (though perhaps highly skilled) intel-

---

[1]For a concise summary of attempts to define creativity within a variety of intellectual traditions, see Carl R. Hausman, 1998, "Creativity: Conceptual and Historical Overview," pp. 453-456 in *Encyclopedia of Aesthetics*, Vol. 1, Michael Kelly, ed., Oxford University Press, New York.

lectual tasks, such as editing manuscripts for spelling and grammar, or applying known techniques for deriving solutions to given mathematical problems. Less obviously, creativity can be distinguished from innovation; there are plenty of software products and business plans that are (or were) innovative, in the sense of accomplishing something that had not been attempted before, without being particularly creative. And there are many original scientific ideas that turn out to be wrong. There is always something unexpected, compelling, and even disturbing about genuinely creative production.[2] It claims value, and it has an edge. It challenges our assumptions, forces us to frame issues in fresh ways, allows us to see new intellectual and cultural possibilities, and (according to Kant, at least) establishes standards by which future work will be judged.[3]

The implicit and explicit ambitions reflected in creative production tend to differentiate it from routine production. It often focuses on unexpected questions rather than those that have already entered the intellectual mainstream. It goes for high payoffs and is undeterred by accompanying high risks. It seeks out big questions rather than opportunities to make incremental advances, and it looks for fundamental change. It is not bothered by rule breaking, boundary crossing, and troublemaking. And it is characteristically reflexive—engaged in reflecting upon and rethinking processes, not just applying them.

Creative production is not always positive and widely valued; one can be creatively evil, and one can waste creative talents on crazy projects that nobody cares about. But the products of creative science, scholarship, engineering, art, and design—even creative basketball—can bring immense benefits to society, as well as providing deep satisfaction to their originators. So respect is accorded to creative individuals and institutions, and society is often willing to invest in projects and programs that plausibly promise (though can never quite guarantee) creative results.

For Plato, and later for the Romantics, creativity was an ineffable attribute of certain mysteriously favored individuals—a gift of the gods. You could cultivate and exercise it if you had it, but there wasn't much else you could do about it. Today's consensus (endorsed by this Committee on Information Technology and Creativity) favors the view that creativity can be developed through education and opportunity,

> There is always
> something
> unexpected,
> compelling,
> and even
> disturbing
> about genuinely
> creative
> production.

---

[2]Hausman ("Creativity," 1998, p. 454) puts the point more technically, as follows: "Not just anything brought into being invites us to call it a creation, however. There is a stronger or radical and normative expectation that what is brought into being regarded as having *newness* and (at least for the creator) *value*. The newness of the outcome of such a radical creative act is a characteristic not simply of another instance of a known class—a numerical newness, such as may be attributed to a freshly stamped penny or a blade of grass that has just matured—but an instance of some new kind. It is a thing that is one of its kind that occurs for the first time, and being thus newly intelligible, is valuable."

[3]Immanuel Kant, 1781, *Critique of Pure Reason*. See <http://www.arts.cuhk.edu.hk/Philosophy/Kant/cpr>. Note that as this report went to press, all URLS cited were active and accessible.

that it can be an attribute of teams and groups as well as individuals, and that its social, cultural, and technological contexts matter. The committee tends to believe that it is possible to identify and establish the conditions necessary for creativity, and conversely, that we risk stifling creativity if we get those conditions wrong. Renaissance Florence clearly provided the conditions for extraordinary artistic and scientific creativity, but it is easy to name many modern cities (we will avoid getting ourselves into trouble by doing so) that apparently do not.

More precisely, creative practices—practices of inquiry and production that seek more than routine outputs and aim instead for innovative and creative results—can be encouraged and supported in some very concrete and specific ways. Society can try to provide the tools, working environments, educational preparation, intellectual property arrangements, funding, incentives, and other conditions necessary to support creative practices in various fields.

## DOMAINS AND BENEFITS OF CREATIVITY

No intellectual domain or economic sector has a monopoly on creativity; it manifests itself (often unpredictably) in multiple fields and contexts. But the manifestations vary in form and character, in associated terminology, and in the types of benefits that result.

In science and mathematics, the most fundamental outcome of creative intellectual effort is important new knowledge. Generally, scientists and mathematicians are clear on the difference between such knowledge and that which results from incremental advances within established intellectual frameworks. Ground-breaking discovery is widely (though not universally) regarded as a product of great value in itself, but it is also valued more pragmatically—as an enabler of technological innovation.

In engineering, and in technology-based industry, creativity yields technological inventions. Such inventions can result in commercially successful products, in improvements to the quality of life (as, for example, when motion picture technology enabled a new form of entertainment, or when an innovative new drug provides a cure for a disease), and in the generation of income streams through intellectual property licensing arrangements. Thus the social and economic benefits are often clearly identifiable and measurable. In recent decades, information technology has been a particular locus of technological invention, the benefits of which need no elaboration here.[4]

---

[4]For an elaboration, see Computer Science and Telecommunications Board, National Research Council, 1999, *Funding a Revolution: Government Support for Computing Research*, National Academy Press, Washington, D.C.

An important manifestation of economic creativity is entrepreneurship—bringing together ideas, talent, and capital in innovative ways to create and make available products and services. Often, in fields such as information technology and biotechnology, close alliances emerge between the institutions of technological innovation (e.g., research universities) and entrepreneurial activity; each one requires and motivates the other. This is particularly evident in fast-moving, high-tech economic clusters, such as the information technology cluster in Silicon Valley or the biotechnology cluster of Cambridge, Massachusetts.

Cultural creativity manifests itself in the production of works of art, design, and scholarship. Like contributions to scientific and mathematical knowledge, such works are highly valued in themselves. Nations and cities take immense pride in their major cultural figures, their cultural institutions, and their cultural heritage. Many value the experience of producing as well as consuming art, design, and scholarship. Not only high cultural practices, such as opera at the Metropolitan in New York City, but also popular practices, such as amateur photography, may be valued for the participant experiences they provide.

Practices of cultural creativity also provide the foundation of the so-called creative industries that seek profits from production, distribution, and licensing.[5] One component of the creative industries consists of economic activity directly related to the world of the arts—in particular, the visual arts, the performing arts, literature and publishing, photography, crafts, libraries, museums, galleries, archives, heritage sites, and arts festivals. A second component consists of activity related to electronic and other newer media—notably broadcast, film and television, recorded music, and software and digital media. And a third component consists of design-related activities, such as architecture, interior and landscape design, product design, graphics and communication design, and fashion.

There are some problems with the very idea of "creative" industries. Creativity clearly is not confined to them, and much of what they engage in could hardly be called creative in any sense. Sometimes, as when they devote their efforts to churning out routine "content," they even seem actively counter-creative. Still, the creative industries do ultimately depend on talented, original artists, designers, and performers to create the value that they add to and deliver, while many artists, designers, and performers depend on the infrastructure of the creative industries and are rewarded by their engagement with the creative industries. The idea is problematic in some respects and there-

It is possible to identify and establish the conditions necessary for creativity, and conversely, we risk stifling creativity if we get those conditions wrong.

---

[5]The U.K. Creative Industries Taskforce, in its 1998 report *Creative Industries Mapping Document*, defined the creative industries as "those industries which have their origin in individual creativity, skill and talent and which have the potential for wealth and job creation through the generation and exploitation of intellectual property." See <http://www.culture.gov.uk/creative/creative_industries.html>.

fore should be treated with appropriate critical caution, but it remains a useful one. And, in any case, the name "creative industries" has stuck.

## THE CREATIVE INDUSTRIES

That the creative industries are now big business hardly needs emphasis. There have been numerous recent efforts to quantify this intuition by measuring their economic contributions. According to an estimate developed by Singapore's governmental Workgroup on Creative Industries,[6] the United States led the way in the creative industries in 2001, with the creative industries accounting for 7.75 percent of the gross domestic product (GDP), providing 5.9 percent of national employment, and generating $88.97 billion in exports,[7] with the United Kingdom, Australia, and Singapore also exhibiting industry sectors of significant size (representing 5.0 percent, 3.3 percent, and 2.8 percent of national GDP, respectively).[8]

The United States has some important, major creative industry clusters,[9] notably those of Los Angeles (with a particular emphasis on film, television, and music), New York (with a particular emphasis on publishing and the visual arts), San Francisco (with particular recent emphasis on digital multimedia), and some smaller, more specialized clusters in cities such as Boston, Austin, and Nashville. In Europe, many creative industry clusters, such as those of London, Paris, and Milan, have developed at long-established centers of culture. In Australia, significant new clusters, based mostly on film, television, and digital multimedia, have emerged in Sydney, Melbourne, and Adelaide. A cluster oriented toward software tools and production has developed in Canada, especially in Toronto and Montréal. The value of such clusters is obvious, so it is not surprising that there has been growing worldwide interest in the regional development strategy of

---

[6]Workgroup on Creative Industries, 2002, *Creative Industries Development Strategy: Propelling Singapore's Creative Economy*, Singapore, September, p. 5.

[7]For core copyright industries only.

[8]These numbers represent the percentages of GDP for different years between 1997 and 2001, though the fundamental point remains valid—the creative industries represent a significant segment in each nation's economy. See Economic Review Committee, Government of Singapore, 2002, "The Rise of the Creative Cluster," *Creative Industries Development Strategy*, p. 5. Available online at <http://www.erc.gov.sg/pdf/ERC_SVS_CRE_Chapter1.pdf>.

[9]Industry clusters are often defined as "concentrations of competing, collaborating and interdependent companies and institutions which are connected by a system of market and non-market links" (see definition of the Department of Trade and Industry, United Kingdom, available online at <http://www.dti.gov.uk/clusters/>).

encouraging creative industry clusters.[10]  In the United Kingdom, for example, each of the ten Regional Development Agencies has focused on the creative industries as a growth sector, and each local authority is mandated by the Department of Culture, Media and Sport to produce a development strategy for the creative industries.[11]

It is important to recognize that these creative clusters do not just consist of large firms. They also encompass independent artists and designers, numerous small businesses, cultural institutions such as galleries and performing arts centers, and educational institutions. Many of those involved with the creative industries may play multiple roles; for example, artists and designers may combine independent practice with teaching, and employees of large firms may "moonlight" with small practices of their own.

But the creative industries also have a strategic importance that extends beyond regional economic development. In a progressively interdependent world where culture tempers and inflames politics as well as markets, strong creative industries are a strategic asset to a nation; the predominance of Hollywood movies, Japanese video games, and Swiss administration of FIFA soccer are forms of soft power that have global, albeit subtle, effects, particularly in countries whose bulging youth populations have access to television and the Internet. Movies, music videos, fashion, and design foster aspirations in the developed and developing world. It matters that teenagers in China—or Pakistan—idolize Michael Jordan. The ability to generate a cultural agenda via the arts, design, or media is a form of deep, pervasive influence and is as integral to global leadership as trade policy or diplomatic relationships.[12]  Globally available cultural products serve as a kind of common social currency in an increasingly fractured and fractious world. To that extent, the reach and robustness of a nation's creative practices can constitute a form of global leadership—while also, of course, potentially attracting charges of cultural imperialism. A nation's creative practices can also provide valuable visibility and branding, as with Italian and Finnish design.

Creative clusters

encompass

independent

artists and

designers,

numerous small

businesses,

cultural

institutions such

as galleries and

performing arts

centers, and

educational

institutions.

---

[10]See Chapter 7 for further discussion.

[11]See Department of Culture, Media and Sport, Creative Industries Program, 2000, *Creative Industries: The Regional Dimension*, Report of the Regional Issues Working Group, February. Available online at <http://www.culture.gov.uk/creative/index.html>.

[12]See Shalini Venturelli, 2001, *From the Information Economy to the Creative Economy: Moving Culture to the Center of International Public Policy*, Center for Arts and Culture, Washington, D.C., available online at <http://www.culturalpolicy.org/pdf/venturelli.pdf>; and Joseph S. Nye, 1999, "The Challenge of Soft Power:  The Propounder of This Novel Concept Looks at Lloyd Axworthy's Diplomacy," *Time*, February 22, available online at <http://www.time.com/time/magazine/intl/article/0,9171,1107990222-21163,00.html>.

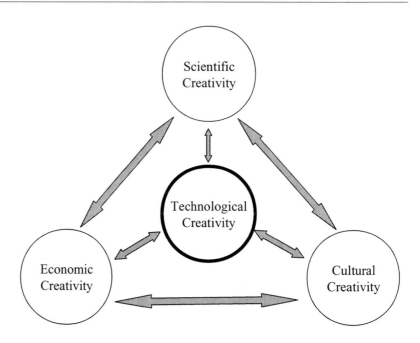

FIGURE 1.1 Domains of creative activity.

## INTERACTIONS AMONG DOMAINS OF CREATIVE ACTIVITY

For some purposes it is useful to distinguish scientific, technological, economic, and cultural creativity, as discussed above. But it is also important to emphasize that these domains are often tightly coupled, and that activity in one may depend on parallel activities in others. This committee was charged and constituted to focus primarily on creative practices in the arts and design, and their intersections with information technology, but it recognized that the coupling to other domains of creative practice could not be ignored.

Figure 1.1 illustrates the approximate nature of that coupling—each broad domain of creative practice in two-way interaction with every other. Reflecting a National Academies/Computer Science and Telecommunications Board perspective, technological creativity is shown at the center of the diagram, which of course could also be redrawn to show any of the other domains at the center.

Consider, for example, some important interactions of technological creativity with other domains. Scientific discovery sometimes drives technological invention, but conversely, the pursuit of technological innovation often suggests scientific questions and ideas. Similarly, entrepreneurial energy may motivate engineering and product innovations, but such innovations may also demand creative strategies for successfully bringing inventions to the market. And newly invented technologies may produce bursts of artistic and design creativity—as with Renaissance perspective, photography, film, radio, television, and computer graphics—while the work of artists and designers may generate desire for technological innovations, shape the directions of technological investigations, and provide critical perspectives.

The additional reciprocal relationships indicated by Figure 1.1 are no less worthy of note. In the creative industries, innovative entrepreneurs develop new ways to produce and distribute creative products, while creative production often demands of businesses and institutions (such as museums, cultural foundations, and art and design schools) new distribution, curatorial, preservation, and other strategies. There is a subtle, complex, but undoubtedly important (in some periods, at least) relationship between the intellectual frontiers of the creative arts and the sciences. And there are even cross-relationships between scientific and economic innovation—as when physics Ph.D.s moved to Wall Street and brought with them new tools and methods for the financial industry.[13]

Innovative design is often situated precisely at the intersection of technologically and culturally creative practices. On the one hand, designers are frequently avid to exploit technological advances and to explore their human potential. On the other, they typically have close intellectual alliances with visual and other artists. And innovative design can yield high economic payoffs; firms such as Apple, Sony, Audi, and Target have differentiated themselves and in some cases turned themselves around through innovative design. Volkswagen remade its image, and refreshed its reputation for witty innovation, with the revived and redesigned Beetle. Bilbao put itself on the world map by building the highly innovative Bilbao Guggenheim Museum—a work that embodies many technological innovations and at the same time is engaged with the frontiers of the visual arts. South Korea has recently had great success with a national policy of emphasizing quality and innovation in the design of consumer products.

These various interrelationships suggest the importance not only of specialized loci of creativity, such as highly focused research laboratories and individual artist's studios, but also of creativity clusters—complexes of interconnected activity, encompassing multiple domains, which provide opportunities and incentives for productive cross-fertilization. Thus a laboratory director might seek to establish a creative, cross-disciplinary cluster of individuals and research groups at the scale of a small organization; a research university provost might seek a creative cluster of departments, laboratories, and centers at the scale

> Innovative design is often situated precisely at the intersection of technologically and culturally creative practices.

---

[13]See, for example, "Physicists Graduate from Wall Street," *The Industrial Physicist*, December 1999, available online at <http://www.aip.org/tip/INPHFA/vol-5/iss-6/p9.pdf>.

of a campus; regional planners might try to encourage formation of creative industrial and institutional clusters within their jurisdictions; and national strategists might seek to do so at even larger scales.

The importance of such efforts is increasingly widely recognized, and a related research and policy literature is emerging: Economists explore the proposition that "economic growth springs from better recipes, not just from more cooking"—that is, from the generation and application of innovative and creative ideas;[14] planners analyze "creative cities" and "creative regions" that attract and retain talent, and that provide environments in which creative practices flourish;[15] the idea of a "creative class" has quickly become popular;[16] and the possibility of shifting from "the information economy" to "the creative economy" has become a hot topic among policy makers from Scotland to Hong Kong.[17]

## THE ROLES OF INFORMATION TECHNOLOGY

Figure 1.2 shows a more specialized version of Figure 1.1, in which information technology replaces technological creativity at the center. Information technology has important relationships to creative practices in other domains. It benefits enormously from basic scientific and mathematical advances, and in return, it provides scientists and mathematicians with powerful new tools and methods. It provides entrepreneurs with a stream of opportunities to develop and market new products and services, while benefiting from the research and development investment that the prospect of successful commercialization motivates. It provides artists and designers with whole new fields of creative practice, such as computer music and digital imaging, together with tools for pursuing their practices in both new and established fields, while benefiting from the inventive and critical insights that artists and designers can bring to it. Increasingly, the committee suggests, information technology constitutes the glue that holds clusters of creative activity together.

The effectiveness of information technology as glue is enhanced by its extraordinary capacity to apply the same concepts and techniques across many different fields. Once content is reduced to bits, it doesn't much matter whether it represents text, music, scanned im-

---

[14]See Paul M. Romer, 1993, "Economic Growth," *The Fortune Encyclopedia of Economics*, David R. Henderson, ed., Warner Books, New York.

[15]See Charles Landry, 2000, *The Creative City: A Toolkit for Urban Innovators*, Earthscan, London.

[16]See Richard Florida, 2002, *The Rise of the Creative Class*, Basic Books, New York.

[17]See John Howkins, 2002, *The Creative Economy: How People Make Money from Ideas*, Penguin Books, London.

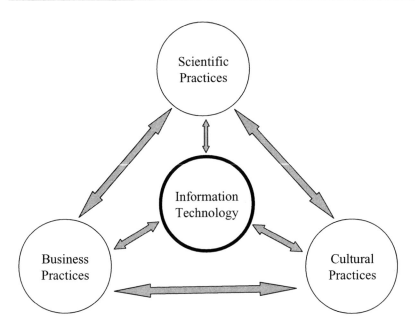

FIGURE 1.2 Information technology as glue.

ages, three-dimensional (3D) computer-aided design (CAD) models, video, or scientific data; the same techniques, devices, and channels can be used for storing and transporting it. Once you have a stream of digital data, whatever the source, you can apply the same techniques to process it. Once you have an efficient sorting algorithm, you can use it to order vast files of scientific data or to arrange polygons (digital objects) for hidden-surface removal in rendering a 3D scene. Once you have a library of software objects, you can use those objects as building blocks to quickly construct specialized software tools for use in many different domains.

Furthermore, information technology can support the formation of non-geographic clusters of creative activity. In the past, such clusters depended heavily on geographic proximity for the intense face-to-face interaction and high-volume information transfer that they required. (If you were in the movie business you wanted to be in Hollywood, if you were in publishing you wanted to be in New York, and so on.) Distance is not dead, and these things still matter, but efficient digital telecommunication now supports new types of clusters. Architectural projects, for example, are now routinely carried out by geographically distributed team members who exchange CAD files over the Internet and meet by videoconferencing—with the advantage that specialized talent and expertise can be drawn from a global rather than local pool. Long-distance electronic linkages between local clusters, such as that between the film production cluster in Hollywood and the postproduction cluster in London's Soho, are also becoming increasingly important.

And the growing integration of digital storage and processing technology with networking technology and sensor technology is even further strengthening the role of information technology as glue. In many domains, cycles of production, distribution, and consumption can now be end-to-end digital. For example, the production and distribution of a photographic image used to entail a silver-based chemical process for capture, half-toning followed by a mechanical process to print large numbers of copies, and physical transportation to distribute those copies; now, capture can be accomplished by means of a charge-coupled device (CCD) array in a digital camera, replication becomes a matter of applying software to copy a digital file, and distribution through the global digital network can follow instantly.[18]

Finally, many argue that information technology is, by its very nature, a powerful amplifier of creative practices. Because software can readily be copied and disseminated, and because there can be an unlimited number of simultaneous users, software supports the dissemination, application, and creative recombination of innovations on a massive scale—provided, of course, that intellectual property arrangements do not unduly inhibit creative work. Much of the current debate about intellectual property and information technology focuses on questions of how best to support, encourage, and reward creative practices.[19]

In summary, information technology now plays a critical role in the formation and ongoing competitiveness of clusters of creative activity—both geographic clusters and more distributed clusters held together by electronic interconnection and interaction. IT is an important driver of the expanding creative industries. And, due to several factors, its role as glue is strengthening. First, the generalizability of digital tools and techniques across multiple domains makes them particularly efficient and effective in this role; they can displace predigital tools and techniques, as in the cases of CAD displacing drawing boards and drafting instruments and digital imaging displacing silver-based photography. Second, the increasingly effective integration of diverse digital technologies is producing efficient, large-scale, multipurpose production and distribution systems that can effectively serve the creative industries. Third, these systems support the formation of non-geographic clusters of creativity that can draw on global talent pools. And finally, the amplification effects that are inherent to information technology are likely to have strong (sometimes unexpected) multiplier effects; they may unleash waves of scientific and mathematical, technological, economic, artistic, and cultural creativity.

*Many argue that information technology is, by its very nature, a powerful amplifier of creative practices.*

---

[18]CCD arrays consist of tiny light sensors that encode scenes as sets of intensity values. The larger the array, the finer the resolution of the picture. The development of digital cameras has been driven, generation by generation, by the release of successively larger CCD arrays.

[19]See Chapter 7 for an articulation of the important role played by intellectual property issues in information technology and creative practices.

• • • • • • • • • • • • • • • • • • • • • • • • • • • • • • •

# THE RACE FOR CREATIVITY IN A NETWORKED WORLD

It seems to this committee that there is an emerging, global race to establish effective, sustainable clusters of IT-enabled creative activity at local, regional, and national scales—and at even larger scales, like that of the European Union. A number of studies and initiatives are directed at this goal, such as the Seoul Digital Media City project (South Korea),[20] the BRIDGES International Consortium on Collaboration in Art and Technology (Canada),[21] the Massachusetts Museum of Contemporary Art (United States), the Kitchen's national art and technology network (United States),[22] and the National Endowment for Science, Technology and the Arts (United Kingdom).[23] The rewards are high; such clusters are engines of economic growth, of enhanced quality of life, and of cultural and political influence—that is, of soft power. Success in launching and sustaining them depends on capacity to attract and retain creative talent, on establishing the conditions and incentives necessary for that talent to flourish, and—increasingly—on the effective exploitation of information technology.

In the following pages, the committee focuses specifically on clusters of creative activity in the arts and design and their interactions with information technology, as illustrated in Figure 1.3—which is simply a subset, but a crucial one, of Figure 1.2. The interactions between these two domains are important not only for their mutually beneficial effects, but also because they help to energize larger systems of interconnected creative activity. This report provides more detailed analyses of the conditions needed for creativity in a networked world and recommends strategies for establishing and sustaining successful clusters of IT-related creative activity in the arts and design. It asks the following questions:

1. How can information technology open up new domains of art and design practice and enable new types of works?

2. How can art and design raise important new questions for information technology and help to push forward research and product development agendas in computer science and information technology?

---

[20]See <http://www.dmc.seoul.kr/english/why/overview.jsp>.

[21]See <http://www.banffcentre.ca/bnmi/bridges/>.

[22]See <http://www.thekitchen.org>.

[23]Additional information about the Massachusetts Museum of Contemporary Art and other initiatives is given in Chapters 7 and 8.

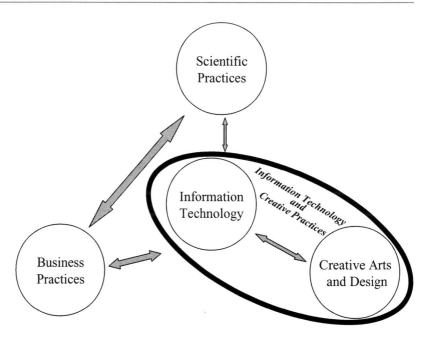

FIGURE 1.3 Information technology and creative practices.

3. How can successful collaborations of artists, designers, and information technologists be established?

4. How can universities, research laboratories, corporations, museums, arts groups, and other organizations best encourage and support work at the intersections of the arts, design, and information technology?

5. What are the effects on information technology and creative practices work of institutional constraints and incentives, such as intellectual property arrangements, funding policies and strategies, archiving, preservation and access systems, and validation and recognition systems?

## ROADMAP FOR THIS REPORT

This chapter provides an introduction to the world of information technology and creative practices (ITCP) and outlines the benefits to the economy and society from encouraging and supporting work in this new domain. Chapter 2 explores the systemic nature of creativity and how multiskilled individuals and collaborative groups pursue work. Various factors, from differences in communication style and vocabulary to evolving work environments, influence how this work is carried out. Information technology is the focus of Chapters 3 and 4.

Chapter 3 analyzes the role of IT in supporting ITCP work and offers observations for the future design of improved IT tools that would provide better support of ITCP. This role for IT—in service to other disciplines—is a well-appreciated one. Chapter 4 challenges this traditional role to consider how the art and design influences of ITCP can help to advance the discipline of computer science. Chapters 5 and 6 discuss the venues for conducting, supporting, and displaying ITCP work. A wide range of venues—from specialized centers for ITCP to museums and corporations—is explored in Chapter 5; ITCP-related programs and curricula in schools, colleges, and universities are covered in Chapter 6. Institutional and policy issues such as intellectual property concerns, digital archiving and preservation, validation and recognition structures, and regional planning are presented in Chapter 7. In Chapter 8, the policies and practices for funding ITCP work are described and analyzed. The discussions from the chapters are synthesized and findings and recommendations are articulated in the report's opening "Summary and Recommendations" chapter. Although these findings and recommendations are directed to particular decision makers such as university administrators, officers of funding agencies, or directors of cultural institutions, many of the ideas are applicable to multiple decision makers, given that ITCP transcends current institutional, disciplinary, and professional boundaries.

# 2 | Creative Practices

P eople are the engines of creative practice. To work within the realm of information technology and creative practices (ITCP), individuals or groups need to be fluent in multiple disciplines. Some individuals can simultaneously master multiple subject domains (modern-day Leonardo da Vincis) as required, whereas others participate in collaborative groups of people with complementary or synergistic expertise and skills. Each approach presents its own set of advantages and challenges. Each approach also benefits from resources such as training tools and suitable working conditions. This chapter explores how human creative capabilities can be accessed, developed, and applied to ITCP work.

The first section briefly reviews what it means to be fluent in the ITCP context and outlines the role that individuals and groups with such abilities play in producing ITCP work. The second section discusses, and explores how institutions might enhance, the two basic approaches to work in ITCP: individuals alone (e.g., an independent artist), and collaborative groups of various types (e.g., a team developing a video game). The third section discusses key challenges that arise in cross-disciplinary collaborations. The final section outlines resources that can support the human capability to create meaningfully.

## WHAT MAKES PEOPLE CREATIVE

What makes one action ordinary and another creative? Part of the answer is personality, although there has been surprisingly little study of creativity by psychologists.[1] Research points to a tendency for

---

[1]A survey of psychological research on creativity, intended to motivate more attention, can be found in Dean Keith Simonton, 2000, "Creativity: Cognitive, Personal, Developmental, and Social Aspects," *American Psychologist* 55(1): 151-158.

creative people "to be independent, nonconformist, unconventional, even bohemian, and . . . to have wide interests, greater openness to new experiences, a more conspicuous behavioral and cognitive flexibility, and more risk-taking boldness."[2] Part of the answer is behavioral, including the extent to which deliberation and skill are involved. Deliberation involves making choices about things that matter. "Fasting," Nobel laureate Amartya Sen has famously written, "is not the same thing as being forced to starve. Having the option of eating makes fasting what it is: choosing not to eat when one could have eaten."[3] Other factors relate to context, such as the nature of one's experiences, notably "(a) diversifying experiences that help weaken the constraints imposed by conventional socialization and (b) challenging experiences that help strengthen a person's capacity to persevere in the face of obstacles"[4]—both of which are characteristic of an emergent field in general and ITCP in particular. Interestingly, a factor in achieving diversifying and challenging experiences may be cultural diversity; there is evidence that exposing a culture to alien influences and experiencing marginality or even dissent are correlated with creativity.[5] More generally, the start of a creative act is the escape from one range of assumptions—a context—often with the aid of another context seemingly at odds with the first but that provides a new way of viewing what we already thought we understood. The arts do this for IT, and IT does this for the arts.[6]

Creativity can be linked to tools, which have been a constant factor in the arts as well as in science and engineering. Because ITCP is defined with reference to a set of tools—IT—it calls for an understanding of creativity as human complements to digital capabilities: the opportunity, knowledge, and skill to make disciplined judgments about how and when to use or not use those capabilities. Although novices can now enter many fields through interfaces—provided by software packages—that encapsulate and parameterize aspects of specialized trades and crafts that previously took lifetimes to learn, learning to use a tool does not of itself make one a skilled practitioner.

There is a difference between basic functional know-how (e.g., knowing a few words of a foreign language) and higher-level skill, or

---

[2]Simonton, 2000, "Creativity," p. 153.

[3]Amartya Sen, 1999, *Development as Freedom*, Alfred A. Knopf, New York, p. 75. The committee is indebted to Mansell (2001) and Garnham (1997) for their readings of Sen in terms of communication and media policy. See Robin Mansell, 2001, "New Media and the Power of Networks," First Dixons Public Lecture, London, October 23, available online at <http://www.lse.ac.uk/Depts/Media/rmlecture.pdf>; and Nicholas Garnham, 1997, "Amartya Sen's 'Capabilities' Approach to the Evaluation of Welfare: Its Application to Communications," *Javnost-The Public* 4(4): 25-34.

[4]Simonton, 2000, "Creativity," p. 153.

[5]Simonton, 2000, "Creativity," p. 155.

[6]Allucquère Rosanne Stone, in *The War of Desire and Technology at the Close of the Mechanical Age* (MIT Press, Cambridge, Mass., 1995), describes how technology can provide prostheses, expanding and enhancing one's interaction with the world.

fluency.[7]  Previous studies of fluency in the use of IT have distinguished between general intellectual capabilities; IT-specific but device-independent generic concepts; and a more contingent class of specific, device-dependent technical skills. In *Being Fluent with Information Technology*,[8] important generic IT conceptual capabilities are identified, including algorithmic thinking, facility with principles of knowledge representation, and adaptability to change.[9]  These conceptual capabilities represent a level of understanding that goes far beyond how to use a given software package. Relatively few artists may pursue true IT fluency, since artists usually learn what they need to know, appropriate the necessary technology and materials, and make their art, but some movement in that direction appears important for ITCP.[10]

Early ITCP has been associated with artists' frustrations with IT, and ease of use for non-technically expert or non-fluent artists and designers is a concern. Yet highly creative performance by artists and designers has been associated with tools that are somewhat difficult to use,[11] especially when the alternative is ease of use achieved through preprogrammed and therefore limiting or constraining features. Creative people always struggle against the limits of their medium—wood splits, musical instruments have limits to pitch and volume, and so on. The challenges presented by IT have helped to stimulate some kinds of art and design—and artists' responses to those challenges, from seeking better tools to exploiting the flaws in or breaking those available as part of their art, should help to stimulate development of new forms of IT.

---

Relatively few

artists may

pursue true IT

fluency, but

some movement

in that direction

appears

important

for ITCP.

---

[7]An emphasis on individual talent and resourcefulness is, of course, commonplace in the traditional arts. A novice musician can pick up an instrument and make sounds. Skilled musicians, though, can make bad instruments sound sweet, and they alone have the virtuosity to "possess" great ones.

[8]Computer Science and Telecommunications Board, National Research Council, 1999, *Being Fluent with Information Technology*, National Academy Press, Washington, D.C.

[9]Paul David, in a similar vein, discusses the importance of generic learning abilities, which must go beyond the acquisition of a specific repertoire of techniques, or even the ability to cope with a need for constant updating of technical knowledge, to a "capacity to understand and anticipate change." See Paul David and Dominique Foray, 2002, "An Introduction to the Economy of the Knowledge Society," *International Social Science Journal (UNESCO)* 171:9.

[10]When artists try to learn skills for their art they are very well motivated. They see the skills as a way to do an excellent job, to do exciting work (they like the results), and to distinguish themselves from other artists. They may become interested in the intrinsic qualities of the IT, but this is more unusual. (Bill Alschuler, California Institute of the Arts, 2002, personal communication.)

[11]Of course, artists and designers do not like more difficult tools per se. Instead, the committee is acknowledging the usual tradeoff between flexibility and advanced features with preprogrammed solutions and ease of use. For Harold Cohen, artists' tools and instruments have to be "difficult enough to stimulate a sufficient level of creative performance, and you don't do that with something that's easy to use." For further discussion of this point, see Chapter 3 in Pamela McCorduck, 1990, *Aaron's Code: Meta-Art, Artificial Intelligence, and the Work of Harold Cohen*, W.H. Freeman, New York.

When people or groups are fluent in IT and arts and design disciplines, they may work at either of two intersections of information technology and creative practices. The first involves the use of computational technologies as a medium for cultural practices (i.e., viewing IT as providing tools in support of the arts and design fields), stressing the continuities between IT and older technologies and the need for a malleable cultural informatics[12] that remains attuned to traditional practices such as reading, singing, painting, or dancing. The second stresses art as a form of research or knowledge production that is interwoven with the practice of research in IT. There is a lot happening at both intersections, and, despite their superficial differences, the intersections are synergistic and might even be described as flip sides of the same phenomenon. These intersections serve as the bases for the committee's examination in Chapters 3 and 4.

In seeking to understand ITCP and the people who do this work, the committee found it useful to examine not only the content of the work involved, but also the details of how it is organized, both socially and institutionally. As is further discussed in Chapters 5 and 6, distinctive new institutional structures have appeared over the past century, combining studio or atelier creation with research-oriented knowledge production in educational, cultural, scientific, and business contexts. All these institutional contexts attempt to balance and support a variety of interests simultaneously. This hybridity is apparent in the shifting roles individuals play both alone and in teams in such settings, be it as artist, designer, researcher, theoretician, entrepreneur, or technician. A similar hybridity was also evident in the artifacts that the committee considered best to exemplify the intersections of IT and creative practice—rather than material objects, they tended to be processes (e.g., interactive works) with social and material aspects, which span boundaries and can be understood in different ways depending on social context.

These observations correspond closely to the social model of creativity proposed by Mihaly Csikszentmihalyi. In this model, creativity is a three-part social system made up of individuals (or groups of individuals), knowledge domains, and institutional structures. As illustrated in Figure 2.1, *individuals* (or groups) produce new variations on inherited conventions stored in *domains*. These novelties are promoted or filtered in the *field* of social institutions, which select the genres, theories, and technologies that become the new conventions for the continuously updated knowledge domains, and that thus are recycled to form new sources for individual creativity. The field component implies that "colleagues are essential to the realization of indi-

When people or

groups are fluent

in IT and arts and

design disciplines,

they may work at

either of two

intersections of

information

technology and

creative practices.

---

[12]Cultural informatics is "a practice of technical development that includes a deep understanding of the relationship between computer science research and broader culture," according to Phoebe Sengers ("Practices for Machine Culture: A Case Study of Integrating Artificial Intelligence and Cultural Theory," *Surfaces*, Vol. VIII, 1999).

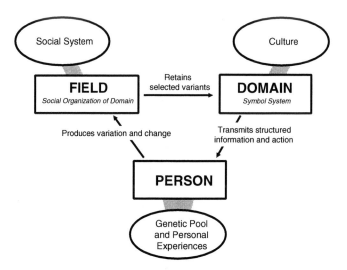

FIGURE 2.1 A systems view of creativity. This map shows the interrelationships of the three systems that jointly determine the development of a creative idea, object, or action. The individual takes information provided by the culture and transforms it, and if the change is deemed valuable by a field, it will be included in the domain, thus providing a new starting point for the next generation of creative persons. The actions of all three systems are necessary for creativity to occur. SOURCE: Derived from Mihaly Csikszentmihalyi, 1987, "A Systems Approach to Creativity," p. 326 in *The Nature of Creativity. Contemporary Psychological Perspectives,* R. Sternberg, ed., Cambridge University Press, Cambridge, U.K.

vidual creativity, . . . because creativity does not exist until those making up the field decide to recognize that a given creative product represents an original contribution to the domain."[13] (See "Validation and Recognition Structures" in Chapter 7.)

Framed in terms of this social model of creativity as a dynamic system connecting people, institutions, and knowledge domains, the creative core common to IT and the arts becomes easier to identify. Creativity results from the interaction of these three systems. And because the systems perspective underscores the importance of a community of practice to sustaining creativity, it also demonstrates the importance of understanding what it means to foster and sustain a community of practice, a goal of this report.

It may be helpful to consider an example. The work of Karim Rashid in industrial design is an illustrative case (see Box 2.1). The boundary-pushing influence of ITCP work on its fields of origin is a recurrent theme in the projects discussed in this chapter.

## HOW CREATIVE PEOPLE WORK

The functional integration of the arts and design fields and IT depends on who is doing what work and how. The human resources

---

[13]Simonton, 2000, "Creativity," p. 155.

## BOX 2.1
### Information Technology and Creative Practices in an Industrial Context

Industrial design has been largely re-created by computer software, from three-dimensional computer-aided design and manufacturing (CAD/CAM) packages to databases that list new grades and alloys of metal and plastic, as well as factories themselves. Products, from automobiles to can openers, have been transformed (it is no accident that consumer goods started getting curvy and ergonomic at around the same time that buildings did). But the process of industrial design has also changed fundamentally, because the time and cost of prototyping have been radically reduced. Designers can generate multiple concept models, honing them in an iterative, evolutionary fashion. But beyond prototyping, information technology has made it possible for industrial designers to engage creatively on a different level—at the level of the manufacturing process itself.

An exemplar of this innovation is Karim Rashid, a highly acclaimed industrial designer whose work includes everything from wastepaper baskets at Bed Bath and Beyond to furniture in the New York Museum of Modern Art's design collection.[1] Rashid's experiments with product manufacturing are possible because modern mass production is increasingly mediated by software. For instance, the apparatus that produced Rashid's curving metal napkin rings for manufacturer Nambé (Figure 2.1.1) is controlled by software that regulates the circumference and length of each napkin ring. By programming the apparatus to vary these parameters randomly, within a range, Rashid was able to create thousands of unique objects, as opposed to thousands of identical objects. The idea of mass-produced one-of-a-kind products—postmodern manufacturing—is possible because one talented individual can bridge the worlds of engineering and consumer aesthetics, and because the technology exists to do so. In the process, the creative professional's role becomes more abstract. It is less about designing objects, and more about designing the process that makes the objects, including the parameters that transcend the designer's direct control. The end result conflates the uniqueness of handcraft with the scale of industrial production.

_____

[1] Further explanation and examples of Rashid's work can be found online at <http://www.karimrashid.com>.

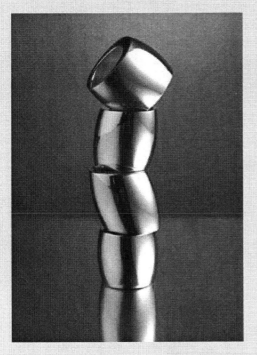

FIGURE 2.1.1 Napkin rings. Photo by Dick Patrick.

can be obtained through the broadening of individual skill sets and through collaborations. Some individuals involved in the arts and design are indeed expanding their knowledge and skills related to IT, and, perhaps less obviously, some computer scientists and engineers are acquiring knowledge and skills in the arts and design (there are distinctive bases in such subdisciplines as graphics and computer music). Collaborations, for all their difficulties, are frequently the preferred and sometimes the required approach, because they demand far less individual investment in learning and therefore accelerate the process of experimentation in combining different kinds of expertise (which is especially important in the early stages of exploration because of the uncertain return from investing time in learning a new area). Another factor arguing for collaboration as a stimulus is the tendency of some cross-disciplinary work, in the absence of a diverse team, to ossify within one discipline or the other. Collaborations may involve anywhere from two to hundreds of people and often are inspired and supported by non-profit organizations or commercial enterprises. Obviously, scale can change the experience and outcome of a collaboration enormously, but while it may seem obvious to suggest that "art," being associated with individuals, requires fewness, the networked nature of modern IT may change that intuition.

## INDIVIDUALS WITH DIVERSE EXPERTISE AND SKILLS

There are some unusually talented people who can do it all, or do enough to create work that straddles more than one discipline and creates new skill sets. This approach has a unique beauty and economy; as one reviewer of this report suggested, an individual's work tends to have a conceptual wholeness, whereas collaborations may produce "camels—horses designed by committees." Many artists prefer the model of the multiskilled individual as the embodiment of the "move fast and travel light" style of work, which allows for a degree of independence in thinking and action that larger collaborative models may not always offer. People who wish to diverge from the political or aesthetic mainstream may want both complete control over their products and independence from external funding and its possible content requirements. Or they may be invested in developing a specific form of personal expression or crafting a concept or theory that they wish to determine independently. These individuals may struggle with the absence of standards in some ways but are able to make their own rules and engage cutting-edge technologies in a personal way to transform an aspect of the world.

There are many models for this style of working, ranging from individuals taking various approaches to the visual arts to novelists to the independent inventor. Growing numbers of artists are becoming skilled in software programming or hardware development, perhaps as a way to maintain a life of the imagination without interference

*Many artists prefer the "move fast and travel light" style of work, which allows for a degree of independence that larger collaborative models may not always offer.*

from a client, patron, or co-worker. The acquisition of such skills is, of course, an implicit acceptance of their value to artistic pursuits. Some people, for instance, can write both computer code and music compositions or turn their code into sculpture. See Box 2.2. Such artists tend to be internally driven by artistic impulses or research interests, although they may benefit from institutional support.

One beneficiary of such support is Michael Mateas, who, at the time he described his work for the committee,[14] was a research fellow in the Studio for Creative Inquiry (an "art think tank") at Carnegie Mellon University (CMU) as well as a doctoral student in computer science.[15] Mateas combines cultural production with artificial intelligence (AI), two activities that normally have very different goals. As he described it, cultural production is interested in poetics (the negotiation of meaning between the artist and audience), artistic abstraction, and audience participation and approval, whereas AI is concerned with task competence, realism, and objectivity. He engages in "expressive AI," building novel architectures, techniques, and approaches. One of his pieces is Terminal Time, an interactive work that constructs documentary videos in real time based on both real historical events and the biases inferred from audience feedback.[16] Terminal Time encompasses a new model of ideological reasoning and a new architecture for story generation, combining the technical capabilities of IT and the dramatic story structure concepts of the arts and humanities in a novel way.[17] In Mateas's view, the project has influenced both the technical research agenda and arts practice—thus fitting nicely into the social model of creativity described above. See Box 2.3.

Individuals who wish to become proficient in multiple fields face at least two formidable challenges. One is the need to deal with enormous and increasing knowledge bases. Trying to remain up-to-date in only one field is demanding enough for most people; the 20th century witnessed tremendous growth in knowledge and a proliferation of disciplinary specialization and narrow professional certification, with a corresponding growth in support structures consisting of professional associations, conferences, periodicals, and curricula. The advent of IT, especially the Internet, has further fueled this trend, especially by facilitating communication among those with niche interests, thus promoting the establishment and maintenance of narrow specialties and interests. Even individuals who already possess both artistic and technical skills may need to learn new ones or find specialists with compatible aesthetic and intellectual views for particular projects.

> Individuals who wish to become proficient in multiple fields face at least two formidable challenges: enormous and increasing knowledge bases . . . and lack of a broad institutional support structure.

---

[14]He briefed the committee at its January 2001 meeting held at Stanford University. See Chapter 6 for further thoughts from Michael Mateas.

[15]As discussed further in Chapter 6, CMU seems to be unusually supportive of cross-disciplinary activities.

[16]For further information about Terminal Time, see <http://www-2.cs.cmu.edu/~michaelm/>.

[17]For further discussion, see Michael Mateas, Steffi Domike, and Paul Vanouse, 1999, "Terminal Time: An Ideologically Biased History Machine," *AISB Quarterly: Special Issue on Creativity in the Arts and Sciences* 102 (Summer/Autumn):36-43.

**BOX 2.2**
## Combining Sculpture, Software, and Hardware Skills

John Simon makes object-based sculptures that combine the skills of painting, sculpture, computer hardware construction, and software development. The work is based on algorithmically generated and intricately cut interfaces between sheets of acrylic plastic—a group of painting-like objects on a wall, with constantly changing patterns on liquid-crystal display (LCD) screens mounted on a structure that is a cross between a painting and a sculpture. See Figure 2.2.1. The software varies the patterns on the screen so that they never repeat. The "painting" is constantly new and constantly changing. Simon's work is in the collection of the Solomon R. Guggenheim Museum and the Print Collection at the New York Public Library.

Simon on his approach and motivation:

> I take the screen and the processor from mostly used laptop computers, which I get from eBay or dealers. I am currently using Apple G3 Powerbooks with 14.1-inch screens. I remove the case and mount the LCD screen to a plastic housing of my own design. The CPU [central processing unit] is mounted on the back of the housing. I install my own software, which runs automatically when the computer is turned on. The images on the screen are constantly changing. This is a way to write software directly for a processor and not have it compete for attention with other things on your desktop. I sell these works through the Sandra Gering Gallery, with which I've had a longtime association. I'm also using a computer-controlled laser to cut and engrave materials like acrylic. I am interested in how the lines and shapes from my algorithmic tools can be manifest in material form.[1]

───────────

[1] See <http://www.creative-capital.org> and <http://www.numeral.com/articles/atkins/decodingdigitalart>.

FIGURE 2.2.1  A work by John Simon. Photo courtesy of John Simon.

**BOX 2.3**

**Terminal Time**

Terminal Time is a mass-audience interactive work that constructs documentary histories in response to audience feedback. The result is similar in style to a Public Broadcasting Service documentary, except that the software constructs the documentary in real time, based on input from the audience. Thus, radically different endings are possible.

The work is produced in the following way. After a 2-minute introduction, the audience is asked three multiple-choice questions. The level of applause from the audience determines the "correct" answer to each question, with the loudest response winning. The answers to this first set of questions are used to create a model of the audience's ideological perspective, which is then used to create a 6-minute video clip representing history from 1000 to 1750 A.D. After the first clip is presented, the process is repeated two additional times, each resulting in the construction of another 6-minute clip, the first representing 1750–1950 and the second 1950–2000. The result is a film constructed in real time. One of the creators of Terminal Time likens it to a genie running amuck, in that the machine infers biases from the audience's responses and then constructs a reinterpretation of history based on exaggerating these biases.

The software running the Terminal Time engine uses an artificial intelligence (AI) architecture consisting of five parts: a knowledge base, a collection of ideology goal trees (goals held by different ideologues), a collection of rhetorical devices (narrative glue for connecting events), a natural language generator, and a media sequencer. Stored in the knowledge base are thousands of terms associated with historical events from the period 1000 to 2000 A.D. Based on an audience's response to each series of questions, the goal trees select historical events from the knowledge base and slant them to accomplish the rhetorical goals of the currently active ideologue. Next, the slanted events are connected together into a story by searching for a sequence of events that can be connected together with the rhetorical devices. The natural language generator then produces the text (based on the connected-together events) that will serve as the voiceover for the documentary. Finally, the system selects and edits together video and audio clips to create the finished documentary. To keep the audience engaged beyond the asking of questions, Terminal Time uses a thematic sequence of rising action, crisis, climax, falling action, and denouement for each complete film.

Cultural productions such as Terminal Time help to synthesize the metaphors of traditional AI, in which the emphasis is often on construction (e.g., to accomplish a particular goal), and art, in which the focus tends to be on conversation (e.g., to create a less-deterministic cultural product). Clearly, this type of cultural production would not be possible without information technology, and computer scientists could not generate this type of content without some knowledge from the arts and humanities concerning how to structure drama.

A second major challenge is the lack of a broad institutional support structure. After decades of experimentation and practice, new hybrid fields are emerging, but with lags in financial support.[18] Limited support, of course, results in limited growth for these particular fields. This is not necessarily bad, because the overall pattern of multiple hybridization results in a number of different intersections of the arts and design with computer science and engineering—different schools of ITCP thought and different kinds of activity—which together span the range from the fine arts through design and craft.

---

[18]See Chapter 6 for an extended discussion.

Multiple hybridization also militates against the institutionalizing of truly creative practice, as any institution formulated to support some particular conception of creative practice will necessarily curtail movement beyond that paradigm (and increasingly, as the institution becomes established).

The growth in cross-disciplinary computing-in-the-arts curricula, which takes many forms, is likely to yield increasing numbers of multiskilled individuals capable of innovating in both technical and artistic/arts-related fields (although at least for a while, their impacts may be concentrated on the arts side, which appears to be more receptive to this type of cross-fertilization than does computer science). Funding may remain a chronic problem, however, for professionals who work (either alone or in groups) outside the commercial sphere.[19] The result is a certain amount of untapped creative energy or underemployment, which limits cultural production to a narrower bandwidth than otherwise might be possible with more generous funding. This situation constrains the breadth and spectrum of the technical syntax of ITCP: **Absent more funding for more experimental work, ITCP may become centered in a commercial, material core.** It may be more pronounced in craft and in design than in art, per se, or fundamental technical research.

> **Absent more funding for more experimental work, ITCP may become centered in a commercial, material core.**

## SUCCESSFUL COLLABORATIONS

Collaborations in ITCP may differ from other kinds of collaborations in that they may well not be symmetrical. Given the differences in training, objectives, and culture, it may be important to articulate different goals between collaborators. A project can be successful and synergistic even if the differing participants have completely different goals for the fruits of the outcome of the collaboration. For example, a particular tool can be used in one way by a scientist and in another way by an artist—but they may develop the tool together (e.g., see the Listening Post project described below). Further, quite different types of relationships between the two communities are possible, each of which embodies different values and therefore requires different techniques in order to achieve success and devise methods for measuring that success.

Collaborations are intense, not superficial, relationships. Less intense forms of relationships include communication (the sharing of information) and cooperation (in which participants influence the decisions of other participants in a common effort). Collaborations may take place in various sizes and forms, ranging from a small project (e.g., academic researchers who agree to work together) to continuing activities within the framework of an institution created for such a purpose (e.g., the studio-laboratories discussed in Chapter 5), to large

---

[19]Because many artistic endeavors are driven by content—artists have their own vision and agenda—rather than profit, they often struggle for support.

commercial enterprises with well-defined and profit-motivated products. What they share is the intention of creating something larger than the sum of their parts. As once noted about the idea of artists working with engineers, "the one-to-one collaboration between two people from different fields always holds the possibility of producing something new and different that neither of them could have done alone."[20]

Non-commercial collaborations often cope with the same inadequate institutional support faced by multiskilled individuals. Such collaborations may often involve people early in their careers who are not yet highly invested in one field nor inhibited by professional norms. They have little to lose by pursuing work that the mainstream might consider marginal. In some cases, radical ideas are the point. The Critical Art Ensemble,[21] for example, is a loosely organized collective of five artists who use "tactical media" to explore the intersections of art, technology, radical politics, and critical theory. Starting out as students looking for a way of organizing that would provide enough financial, hardware, and labor resources to have a cultural impact, the collective now expands and contracts based on specific project needs; the members are geographically diverse and skilled across many disciplines. The results of the work take many shapes—Web sites, performances/installations, and books—which emerge through a horizontal, distributed think-tank process of discussion and exchange among participants. Projects are funded through the participants' "straight jobs," writing and speaking fees, and an occasional sponsor. [22]

Some non-commercial collaborative projects are both inspired and supported by institutions. An example is Bar Code Hotel, an interactive installation by artist/programmer Perry Hoberman that was among nine virtual reality projects produced by the Art and Virtual Environments project at the Banff Centre for the Arts.[23]  In Bar Code Hotel (see Figure 2.2), "guests" enter a room in which the walls are covered with bar codes. The guests use a lightweight wand to activate the black lines in the symbols and issue directives such as "grow" or "fight" to virtual objects they create in a computer—semi-autonomous agents with their own personalities and behaviors. Thus, the guests create a narrative that is partly predetermined and partly spontaneous; when the objects "die" and the guests leave, the hotel returns to

---

[20]From Paul Miller, 1998, "The Engineer as Catalyst: Billy Klüver on Working with Artists," *IEEE Spectrum*, July, available online at <http://www.spectrum.ieee.org/select/0798/kluv.html>. Also see the work of Project Zero at the Graduate School of Education at Harvard University, available online at <http://www.pz.harvard.edu>.

[21]See <http://www.critical-art.net/>.

[22]See <http://www.lumpen.com/magazine/81/critical_art_ensembles.html>.

[23]See <http://www.perryhoberman.com>. Also see M.A. Moser and W.D. MacLeod, eds., 1996, *Immersed in Technology: Art and Virtual Environments*, MIT Press, Cambridge, Mass.

FIGURE 2.2 Bar Code Hotel enables guests to use commands embedded in bar code symbols to interact with semi-autonomous computer-generated objects and create a narrative. Photo courtesy of Perry Hoberman.

its original empty condition. Bar Code Hotel was produced through a hybrid work model in which the artist developed the concept but accomplished the work with help from others. The approach was similar to that of the film business (discussed below in this section) in that it was hierarchical: A producer (Banff) and a director (Hoberman) worked with a team of programmers, sound designers, animators, and other technologists and equipment provided at the Banff Centre.[24] Team members in such situations, while generally carrying out the director's concept, often provide essential ideas.

A small but institutionally driven collaboration, this time involving participants acting as equals, produced the highly successful Listening Post, which monitors online activity in thousands of Internet chat rooms and message boards and then converts these public conversations into a computer-generated opera. This project was instigated and supported by the Brooklyn Academy of Music (BAM), Lucent Technologies' now-defunct pilot program in new media, and by the Rockefeller Foundation. A symposium was set up at which artists and Bell Laboratories engineers and scientists each gave 5-minute presentations on their work and then had the opportunity to talk with each other and find compatible collaborators. Administration was handled by BAM, which awarded $40,000 to each of three

---

[24]For a comparison between Hoberman's directorial role in relation to programmers and other less partitioned team design in the Banff Centre's Art and Virtual Environment's project, see Michael Century and Thierry Bardini, 1999, "Towards a Transformative Set-up: A Case Study of the Art and Virtual Environments Program at the Banff Centre for the Arts," *Leonardo* 32 (4):257-259.

projects. Listening Post was created by two people—Mark Hansen of the Statistics and Data Mining Research Department at Bell Labs, whose cross-disciplinary research draws on numerical analysis, signal processing, and information theory, and Ben Rubin, an artist who works with interactive sound and image technologies. In Listening Post, bits of sampled text are presented as light-emitting diode read-outs and variously pitched speech synthesized to form a screen of visual data accompanied by an "opera" of spoken text. Statistical analysis is used to organize the messages into topic clusters based on their content, tracking the ebb and flow of communication on the Web.

Listening Post demonstrates that collaborations not only draw on and assemble a wide variety of skills in newly developing areas of digital culture but also may alter creative practices themselves—the shape and nature of the way people work, and the way disciplines are defined and categorized. The boundaries of practice here were altered as a result of challenges that arose in the legal territories of intellectual property and licensing. Rubin also explained that, as a result of the collaboration, his "conceptual vocabulary has grown to include notions like clustering, smoothing, outliers, high-dimensional spaces, probability distributions, and other terms that are a routine part of Mark's day-to-day work." He added, "Having glimpsed the world through Mark's eyes, I now hear sounds I would never have thought to listen for." Hansen has expressed similar sentiments, saying: "This installation, its physical presence as well as the underlying intellectual questions, are new for me, as they are for Ben. I suppose it's the mark of a genuine collaboration, that the participants are led in directions they could never have imagined apart."[25]

At the other end of the spectrum of creative work models are larger groupings. Larger groupings tend to be structured according to either the directorial model (which is more common among first-generation media artists) or the low-ego model of distributed responsibility and anonymity (exemplified by the Institute for Applied Autonomy). Many groups have occupied some middle ground between these two. Longer-running collaborations such as Survival Research Laboratories have become a brand with a figurehead, a semi-permanent core, and a tiered and fluctuating membership. Larger, looser groupings occur over the Internet and have their own dynamics, all the way up to large virtual communities.

Some of the more structured and better-funded ITCP collaborations are those found in commercial endeavors, such as segments of the architecture, movie production, and computer game industries. In the film industry, for example, there is a clear hierarchy with well-defined jobs that form a pyramid of synergistic labor to carry out a standardized process of making a product with clearly defined parameters. Such collaborations depend on conventions of practice, standard technologies, and infrastructures for distribution. These are also

> Collaborations not only draw on and assemble a wide variety of skills in newly developing areas of digital culture but also may alter creative practices themselves.

---

[25]See <http://www.earstudio.com>.

professional contexts in which collaborations are the norm, with built-in motivations and rewards for making the process succeed. A shared goal of generating some kind of product or service provides the extrinsic raison d'être for collaboration, communication, and coordination among disparate types of people. The following descriptions of work models in these fields may provide some guidelines for future collaborations of computer scientists and artists and designers.

## Architecture

Architecture is inherently a collaborative field. Only the very smallest design and construction projects are conceived and executed by individuals. Projects of any scale and complexity are undertaken by large teams of specialists—typically including client representatives, architects, specialist engineering consultants, fabricators, subcontractors, and general contractors. The design architect plays a leadership and overall coordination role, taking ultimate responsibility for the quality of a project, but any member of a design and construction team may be called upon to help frame problems and to contribute to their solution. Experienced architectural designers know that innovative, creative projects depend on harnessing the expertise, energy, and imagination of all team members, not just assigning them routine tasks.

Forms of collaboration have evolved as supporting technologies have developed. Medieval architects, for example, were not clearly distinguished from builders, and they spent most of their time on construction sites rather than in separate design offices. Under these conditions, the interactions among team members mostly took the form of on-site, face-to-face discussions, augmented when necessary by the production of simple sketches and full-size templates of detail. With the industrial revolution, a more formalized division of labor emerged:  Architects definitively separated from the construction trades, identified themselves as professionals, increasingly defined themselves as knowledge workers rather than as master craftsmen, and spent most of their time in their off-site ateliers and drawing offices. Drawings on paper became the principal means of developing and recording design ideas, communicating among members of the design and construction team, and establishing construction contracts. Within this new framework, drawings and scale models (rather than on-site construction situations) became the objects of discussion. Collaboration increasingly took place around the drawing board, or in a conference room.

Since the 1960s, digital technology has been transforming design and construction collaboration once again. Computer-aided design (CAD) files have replaced drawings on paper as the primary records of evolving designs. Electronic file transfer and joint access to online databases have increasingly supplanted the physical transportation of drawings as means of communication among design team members.

Videoconferencing and groupware (software tools to support collaboration) play growing roles. As a result, design and construction teams may now be tied together electronically rather than by physical proximity in their interactions and collaborations, they may be distributed geographically, and they may operate asynchronously across multiple time zones.[26] Whereas architecture was once a very local activity, it is now globalizing.[27] Globalization, in this context, means that design and construction teams are not limited to the talent and expertise available locally. They can draw on much larger, more diverse, and competitive talent pools. It is not necessary to go to the structural engineer next door, for example; one can go to a leading international specialist who has exactly the right skills and experience for the current project.

The shift to digital modeling and fabrication based on computer-aided design and manufacturing (CAD/CAM) also provides significantly greater design freedom. Architects can now work, without difficulty, with complex curved surfaces, non-repeating compositions, and other elements that would have been completely unmanageable in the days of hand drafting. And they can use sophisticated software, applied to digital models of projects, to verify structural, thermal, and other aspects of performance. Projects that would have been imaginable but infeasible in the past can now be pursued without much difficulty (see Figure 2.3).[28]

## Movie Production

The movie industry exemplifies cooperative creative practices, relying on collaborative processes involving artists and technicians to make its magic. Temporary task forces of actors, designers, electricians, animators, and many others come together for a single project, working intensely to build relationships and teamwork comparable to that of a string quartet or baseball team.[29] The director may work with writers or composers to develop and revise the screenplay or score, designers and technicians may work together to make the sets, and film editors may rely on digital technologies to create special effects. A

*Architects can now work with complex curved surfaces, non-repeating compositions, and other elements that would have been completely unmanageable in the days of hand drafting.*

---

[26]Similar processes have, of course, unfolded in manufacturing and other contexts where artistic concerns may be less evident (other than in the design component as discussed above).

[27]See Jerzy Wojtowicz, ed., 1995, *Virtual Design Studio*, Hong Kong University Press, Hong Kong; and Jose Pinto Duarte, Joao Bento, and William J. Mitchell, 1999, *The Lisbon Charrette: Remote Collaborative Design*, ISP Press, Lisbon.

[28]See, for example, William J. Mitchell, 1999, "A Tale of Two Cities: Sydney, Bilbao, and the Digital Revolution in Architecture," *Science* 285 (August 6): 839-841; or William J. Mitchell, 2001, "Roll Over Euclid: How Frank Gehry Designs and Builds," pp. 352-364 in *Frank Gehry, Architect*, J. Fiona Fagheb, ed., Abrams, New York.

[29]See Computer Science and Telecommunications Board, National Research Council, 1995, *Keeping the U.S. Computer and Communications Industry Competitive: Convergence of Computing, Communications, and Entertainment*, National Academy Press, Washington, D.C., p. 33.

FIGURE 2.3 Guggenheim Bilbao. Photo courtesy of William J. Mitchell, Massachusetts Institute of Technology.

Hollywood production literally demands this vast array of talent and skill (witness the length and diversity of the credits on a typical film). Then, when the project ends, the team dissolves and the individuals seek new employment elsewhere.[30]  Movie production can also exist on a smaller scale—from the experimental to small-budget independent films. These smaller-scale efforts are also collaborative in nature, with profit or revenue as a less important consideration than it is for mega-Hollywood-scale projects.

Movie production has embraced IT. Indeed, over the past two decades, virtually every facet of movie making has been transformed by IT. Computer-generated imagery (CGI) is commonplace, from the dinosaurs in *Jurassic Park* to the "legless" lieutenant in *Forrest Gump*

---

[30]For economic analysis of the evolution of the formerly dominant studio system to one based as described here, see Richard E. Caves, 2000, *Creative Industries: Contracts Between Art and Commerce*, Harvard University Press, Cambridge, Mass.

and Gollum in *The Lord of the Rings: The Two Towers. Jurassic Park* made history by showcasing the ability to successfully model, render, animate, and composite three-dimensional images at film resolution.[31] Since that film was made, CGI has advanced to the point that, in the words of *Titanic* director James Cameron, "Anything is possible right now, if you throw enough money at it, or enough time."[32] Digital technologies also extend to sound recording, sound production, and picture editing.

The smaller studio and independent film markets also have been transformed by the advent of digital video. The increased scale and portability of cameras have changed shooting styles and are beginning to evolve new aesthetic possibilities. Lower costs are expanding access and the possibilities for experimentation. Thus, niche markets are developing for lower-budget films and are causing an explosion of low-budget production. Desktop tools for postproduction in editing sound as well as animation and special effects are also creating access for a whole new generation of filmmakers. Ironically, as this lower-budget end of film making has achieved commercial viability, it also has tended to compete with the experimental and non-commercial arena of film making for resources, such as access to venues.

Animated work is now being digitized on the scale of feature-length films, as evidenced by the release of *Toy Story* in the mid-1990s. What had been confined to special effects or short demonstrations since the late 1970s has reached a level of maturity able to convince audiences at the subtlest level of expression—character animation, long believed to be beyond the capacity of computer animators. A new Oscar category has been created for "best animated picture"—and the honorees are just as likely (maybe more likely) to be digital artists as traditional cartoonists who draw characters by hand. In fact, many cartoonists are losing their jobs; membership in the screen cartoonists union has dropped by almost 50 percent in the past 5 years.[33] Of course, computer-system animators and cartoonists alike have seen a considerable volume of their work become industrialized, given the division of labor associated with producing a contemporary theatrical film. This does not necessarily spell the end of individual artistry, however, although there is the risk that such artistry is migrating to other realms. Some predict a resilient market for the warmth of traditional animated characters; there may also be new avenues for individual creative practice as the costs of digital workstations fall.[34] Experimentation with short works designed for Web distribution provides

> Desktop tools for postproduction in editing sound as well as animation and special effects are creating access for a whole new generation of filmmakers.

---

[31]Scott McQuire, 1999, "Digital Dialectics: The Paradox of Cinema in a Studio Without Walls," *Historical Journal of Film, Radio and Television*, August, available online at <http://www.findarticles.com/cf_0/m2584/3_19/55610007/p1/article.jhtml?term=+>.

[32]Cited in McQuire, 1999, "Digital Dialectics."

[33]See Claudia Eller and Richard Verrier, 2002, "Animation Gets Oscar Nod as Industry Redefines Itself," *Orlando Sentinel*, February 12.

[34]See Eller and Verrier, 2002, "Animation Gets Oscar Nod."

an outlet for creativity in animation, while the definition of "animation" itself is evolving: Time re-mapping and digital compositing on existing footage extend the notion of animation into territories within film, and in some ways, computer-generated imagery has made all of film into a form of animation.

## Computer Games

Today's game industry, which produces interactive media for personal computers, game consoles (i.e., Playstation 2, Xbox, and Nintendo Game Cube), and online games, is an increasingly important force in youth culture and the economy—video games make more money than the Hollywood box office.[35] Even more than film, computer games require a close marriage between the practical aspects of code and art, and between programmers and artists, at every stage of production. It is not just that different skills are required to produce the end result. Rather, it is the constant state of communication among art, technology, and design that has to be maintained from beginning to end, in order to ship a product.

There are three groups of people involved in the production of a game: designers, programmers, and artists. Designers are responsible for the structure of the experience and the dynamics of interaction between players, or between players and the game world. Programmers are responsible not only for the code that makes this interaction possible, but also for the tools that are used to build the world—unlike film or architecture, most games are built with custom tools because the technology changes so fast. Artists are responsible for the surface of the game—the topography and texture of the world, the way characters look, the animation that occurs when the player takes any kind of action. In the course of production, from concept to completion, these three groups have to work to achieve an almost spousal level of understanding, because their jobs are so interdependent. Designers have to work with programmers to shape the toolkit, to ensure that player interactions will be technically possible. Artists have to talk to programmers, so that they will have enough polygons (or digital objects) to do what they want as well as suitable textural and procedural complexity and character development. Designers and artists must collaborate closely because look and feel are inextricably intertwined. All three groups contribute to the development of game "engines,"[36] which can be reused to develop different games. Game en-

---

[35]According to a report by the NPD Group (as reported in Khanh T.L. Tran, 2002, "U.S. Videogame Industry Posts Record Sales," *Wall Street Journal*, February 7, p. B5), sales of video game software were $9.4 billion in the year 2001, while U.S. box-office receipts totaled an estimated $8.35 billion. Also see Khanh T.L. Tran, 2002, "Consoles Outrun Computers," *Wall Street Journal*, April 19, p. A13.

[36]A game engine supports the basic software elements needed to develop a game, which include rendering, support for sound handling, and other elements and can be reused for other games.

gines have become sufficiently sophisticated that their development is emerging as a category of problems addressed in computer science research.[37]

If the game is played online, all of these groups have to work with a fourth technical group, which oversees the network platform that supports online interaction; this group is responsible for the databases, server arrays, network security, bandwidth allocation, and so forth. Although multiplayer online games may be constrained by network architecture and capabilities, they also may inspire new research and development in these areas. Even something as simple as a player looking through a doorway requires multiple forms of expertise: Can the player see other people outside? If so, that information has to be streamed onto the player's computer—and if there is a crowd outside, performance may suffer. Perhaps there is a way to limit the field of vision (a conversation between programming and design) or compress the graphics files (compromises among art, design, and engineering). Can the other people see the player? (This involves the same issues and more database work.) Instead of segregating tasks, development teams conventionally tackle cross-disciplinary problems by assigning "strike teams," composed of an artist, a programmer, and a designer, to specific problems: artificial intelligence, in-game resources, and so on. High-level, cross-disciplinary collaboration is a daily fact of life. See Figure 2.4.

This level of collaboration exists in part because game technology is a moving target. The medium is evolving so rapidly that many games solve problems that did not even exist a year before, because the tools were not there to solve them. The creation of custom tools to take advantage of leading-edge capabilities means that such teams are working on the edge of what is technically possible, to make a great experience for the player (unlike film, which leverages standardized technologies to a larger degree). Game companies do not have research and development (R&D) departments because every product is a collection of (applied) R&D that eventually has to work, one way or another. In the words of one lead designer, "Every game is a moon shot."[38]

A concept from this industry that may be applicable to other ITCP activities is the leveraging of user talent (not unlike the audience participation in Terminal Time and Bar Code Hotel). The computer game industry is an example of cultural production as a technology

---

[37]Game-engine development has been the focus of doctoral dissertation work at the Naval Postgraduate School's MOVES Institute, for example (personal communication, Michael Zyda, Naval Postgraduate School, March 2002).

[38]Of course, it is worth noting that not every game pushes the technology envelope. Some games, for example, exploit new ideas about social and storytelling approaches that may or may not involve challenging technological problems to solve. And other games may be mostly derivative in nature, using only well-established technology and techniques.

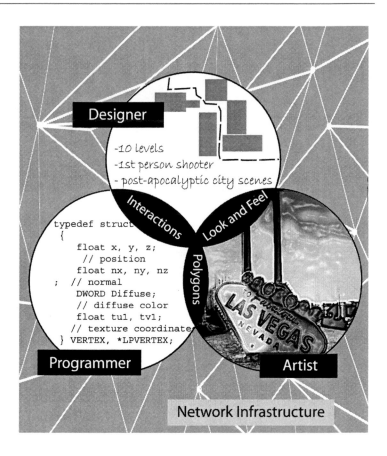

FIGURE 2.4 Development teams for computer games. Illustration created by Jennifer M. Bishop, Computer Science and Telecommunications Board staff.

and market driver, where software engines and authoring tools are regularly made available to end-consumers who use them to redesign or extend the core product, often in directions unanticipated by the publisher. Even something as seemingly reductive as Quake, a first-person shooter, has been reconfigured as a low-tech animation engine—players use the game's editing tools to build environments and characters, which are then manipulated as virtual actors. This is not a market the publisher would have envisioned, much less approached. On the technical side as well, player innovations have driven the artificial intelligence component of the game forward, resulting in smarter code that drives not only sales of the end-product but also commercial licensing of the underlying technology to third-party publishers. Essentially, the flexibility of Quake's tool set has transformed thousands of players into a self-organizing market research and R&D force driven by its own creative imperatives and social incentives.

# CULTURAL CHALLENGES IN CROSS-DISCIPLINARY COLLABORATIONS

Would-be collaborators from different disciplines can encounter a number of obstacles, including difficulties in accessing appropriate funding sources, differences in vocabulary, the absence of frameworks for evaluating non-traditional work, and the long time periods required for projects to gel.[39] Further, it may seem intuitive that the greater the differences between the disciplines involved, the higher these barriers become; one could argue that IT and the creative arts register a high score on this scale. Yet, some claim it is easier to get artists and engineers to work together as a team than it is to get individuals from either group to work with their own colleagues in the same field. That observation has been applied to both computer science and various arts and design fields. Sometimes competition in the same (or a similar) area of expertise is more difficult to deal with than combining different skill sets to attain a common goal.

When adequate resources are available, as is sometimes the case in the corporate world, people can be formally taught skills that are conducive to collaboration. The committee made a site visit to Pixar Animation Studios,[40] a successful company that offers a number of creativity-enhancing activities. Corporate universities, per se, are not new; for example, the Disney Studio offered art classes in its heyday of the 1930s and 1940s.[41] But there is something unusual about Pixar University, a part of the company that has its own "dean" and offers courses in every aspect of filmmaking for Pixar employees (the classes include both technical and artistic "students"). The curriculum includes many forms of studio art (e.g., sculpture, painting, drawing), improvisation, storytelling, and even juggling. Pixar co-founder and president Ed Catmull says a course in improvisation is the closest thing there is to a class in how to collaborate. Perhaps the strongest statement that can be made about these offerings is that they send a signal, coupled with enabling resources and management support, that creativity matters, is encouraged, and may be rewarded, and that it can involve moving beyond one's starting skill set, whether on an individual basis or in combining people with different starting skill sets into teams. In addition, Pixar University contributes to the company's human resources policy by promoting employee retention. Unlike other major studios, Pixar tends to keep its teams together

---

[39]See National Research Council, 2000, *Strengthening the Linkages Between the Sciences and the Mathematical Sciences*, National Academy Press, Washington, D.C.

[40]See Appendix B for a listing of Pixar participants.

[41]See Frank Thomas and Ollie Johnston, 1981, *The Illusion of Life: Disney Animation*, Abbeville Press, New York.

between projects, rather than laying them off. There are downsides, of course, to any strong internal culture, even one designed to promote collaboration and creativity. A self-contained organization without links to external perspectives may encourage homogenous values and an insular view of the world, discouraging the criticism or controversy that often is useful in ITCP work.[42]

Even if specialized training is not an option, general awareness of key issues that arise in collaborations may help projects to succeed. The overall challenge in collaboration is to transcend traditional role boundaries to exploit different perspectives and skills and create new ideas and products that are somehow greater than the sum of their parts. Doing so may involve assessing the multiple dimensions of each relevant discipline—which affects its interfaces to others—and the ongoing processes of change affecting each discipline. Specific obstacles to be overcome at the intersection of IT and creative practices are discussed in the following subsections.

## OVERCOMING PRECONCEIVED NOTIONS ABOUT COMPUTER SCIENTISTS AND ARTISTS AND DESIGNERS

Perceptions about artists and designers and computer scientists can often be formed through popular or anecdotal accounts, rather than through actual encounters. Such perceptions can inhibit mutual respect in collaborations, at least at the outset. The challenge of overcoming such stereotypes permeated the personal accounts of those who briefed the committee and of committee members themselves.[43] Although there are exceptions to and disagreements about stereotypes, some generalizations are useful here for bringing an important issue to light, even at the risk of oversimplification.

Some scientists and engineers exhibit a sense of superiority, if not outright hostility, toward those in the arts and design. Or, put another way, "Artists see science; they don't understand it; they think it is brilliant. Scientists see art; they don't understand it; they think it is dumb."[44] Part of the problem may be the connotations of "creativity" in some contexts. Creativity is often cloaked in an aura of mystery, which suggests that the work results from spontaneous creative insight without rigorous or repeatable methodology, from epiphanies

---

[42]However, companies that wish to keep their work confidential until public release do have reasons for constraining external communication, or at the least, not encouraging it fully.

[43]Of course, such perceptions do not exist in every collaboration. However, testimony to the committee, a review of published literature, and the experiences of most of the committee suggest that the lack of such perceptions is indeed the exception.

[44]Based on discussions at the committee's meeting at Stanford University, January 2001. A reviewer of this report observed that "sentiments here attributed to scientists are seldom encountered among European scientists, probably because U.S. scientists are often unfamiliar with cultural practices."

when alone rather than as a result of sustained discussion with peers; it downplays the analysis, struggle, debate, or committed engagement with pressing social or technical problems. It is very difficult to compare forms of creativity, or sometimes even to recognize them. Some scientists and engineers can also view the arts or other cultural perspectives as luxuries, things that might be supported or pursued as time and resources permit.

Such attitudes may be traceable in part to disparities in funding and, accordingly, some notion of status.[45] The Xerox Palo Alto Research Center (PARC) Artist-in-Residence (PAIR) program, for example, received a certain amount of attention for its attempts to integrate artists with computer scientists and others.[46] Although this program may well have helped Xerox PARC to sustain its creativity, constraints on social integration—accentuated by pay differences—may have limited the creative output.[47] Although people can (and, given discussions within the committee, clearly do) interpret compensation disparities in different ways, national employment statistics show significant differences among workers in the arts and those in technical fields such as computer science; different occupations, even among technical fields, have different earning power, for a variety of reasons that derive from the structure of the economy (and professional conduct).[48] The marked contrast between compensation levels for computer scientists and for artists, other things being equal, is significant for the intersection between IT and the arts inasmuch as it affects collaboration and education. Across organizations, and even departments in a university, compensation levels affect patterns of time use, expectations for research and for infrastructure, and so on.

Similarly, the arts establishment sometimes regards technology suspiciously, as if it lacks a worthy lineage or is too practical to be creative. This attitude was evident in early committee discussions, coming out most strongly in contrasting perspectives on the potential for creative practices within industry. Because of their experience in

---

[45]As Michael Mateas, creator of Terminal Time, told the committee: "Power is a big issue . . . . Certainly in our society there's a power asymmetry between technocrats—scientists and technologists—and artists. Technocrats are . . . in the driver's seat right now in our society."

[46]The context is a research laboratory that had already blended a variety of scientists and engineers and a small group of social scientists.

[47]As characterized to the committee at its January 2001 meeting at Stanford University, the PAIR program when it was launched included "creative" people from the arts with a lot of experience who were paid less than some technical student interns, and who disparaged the scientists as suburban bourgeoisie.

[48]According to economist Richard Caves, creative professionals earn less, on average, than their human capital might suggest, in part because their commitment to producing creative output may lead to different activity and output than would a simpler commitment to satisfying consumers. See Richard E. Caves, 2000, *Creative Industries: Contracts Between Art and Commerce*, Harvard University Press, Cambridge, Mass.; and James Heilbrun and Charles M. Gray, 2001, *The Economics of Art and Culture*, Cambridge University Press, Cambridge, U.K.

deriving research inspiration from practical problems, the technologists found it easier to see creative potential in industry than did the artists, who found more cause for concern about motivations or constraints based on commercial imperatives.[49] Skepticism about technology was also evident in the early days of "Net art" (art using the Internet), which took off in 1994 when the Mosaic browser was first distributed and people realized that the Web was a fertile canvas for art making. Net art was ignored as unimportant at first by art institutions, museums, galleries, art magazines, and funders. (Now that it has gained credibility, some suggest that Internet art may in fact be the medium that best reflects the transformations of the information revolution, the same role that photography and film played in the industrial revolution.[50]) This type of cultural bias can undermine respect and communication, unless the participants are aware of their differences and are willing to modify their behavior appropriately. Although the committee context forced the process of articulating and overcoming such differences among its members, accommodation was neither rapid nor easy, an insight that is important for planning for other contexts.

One concern arising from some quarters of the arts world is that a celebration of the potential of ITCP not become a dirge for more traditional forms of art.[51] One is not a substitute for the other; both should be viewed as complements. Nor should ITCP be viewed as privileging popular forms, such as design, over the fine arts. Although the direct pop culture, because it is so pervasive and so easy to learn and transmit through media, has pushed developed art to the margins, both ends of the spectrum need each other—the direct end to revitalize points of view and connect with basic feelings, the other to reveal much more about an idea (and about ideas) than was first supposed.

The challenge of maintaining respect across disparate fields is an extension of the frequent differences in attitude encountered within a field between researchers in the more theoretical and the more applied areas. More generally, every social context has a prestige and status hierarchy, standards of excellence, standards of language, and modes of expression. It is too late to establish social contexts for ITCP de novo so that everyone is socialized ab initio into shared norms, goals, and expectations. Hence it is important to foster social contexts that recognize explicitly that people come from different cultures and explicitly work to bridge those differences. Establishing strong common goals and simultaneously ensuring individual work satisfaction—the support of individual goals within the group—is one strategy for cross-

---

[49]This perspective is likely to be more common among studio artists than, for example, commercial artists who work in advertising or industrial designers.

[50]Based on a presentation by Mark Tribe to the committee in November 2000 in New York City.

[51]This theme emerged in the review process for this report, for example.

disciplinary communication. Creating an atmosphere of equal value among members is another tactic. Dissension flourishes in an atmosphere of inequity; the collaborative process requires an atmosphere that allows for relaxed exchange.

Perceptions of teamwork in the arts have, in the past, centered on either identical roles (i.e., people working together as equals) or clearly unequal ones (e.g., one person is "in control" and the other is the technician or helper). These models are changing in the wake of new practices such as those used by the Critical Art Ensemble, discussed above. Differentiations between "technicians" and "professionals" shape computer scientists' views of collaborations, too, especially in a cross-disciplinary context. Because people play different roles in teams, assigning credit can be difficult. A major impediment to cross-disciplinary collaborations is the traditional academic focus on isolated disciplines, the organizing principle for departments, journals, and the reward system for teachers and researchers.[52] New technological art forms require new ways of organizing, which can take decades to stabilize, as was true for cinema and perhaps for emergent forms such as virtual environments.[53]

> New technological art forms require new ways of organizing, which can take decades to stabilize, as was true for cinema and perhaps for emergent forms such as virtual environments.

## MINIMIZING COMMUNICATIONS CLASHES

Although the arts and sciences are not completely separate spheres—indeed, some see them as intricately related—they do speak different languages. During the writing of the present report, for example, committee members and staff with IT backgrounds had difficulty understanding the nonlinear concepts and writing style of those with art and critical studies backgrounds. Similarly, a Stanford University computer science professor reported difficulty in collaborating with art historians because they were unfamiliar with data and models.[54] Simply recognizing the barriers posed by jargon, terms of art, and localized practices goes a long way toward bridging such gaps. The Textile Museum in Washington, D.C., for example, took a straightforward approach in demystifying its exhibition of textile art made with digital printing and/or digital weaving techniques, which "allow the artists to investigate traditional textile concepts with a new flexibility and range of creativity."[55] Because casual visitors might have had difficulty understanding either the art pieces or the advantages offered by technology, the museum provided a glossary of textile terms such as "warp" and "weft."[56]

---

[52]See Chapter 6 for a detailed discussion.
[53]See Brenda Laurel, Rachel Strickland, and Rob Tow, 1994, "Placeholder: Landscape and Narrative in a Virtual Environment," *ACM Computer Graphics Quarterly* 28(2):118-127.
[54]Personal communication from Marc Levoy, Stanford University, March 29, 2000.
[55]See "Technology as Catalyst: Textile Artists on the Cutting Edge," 2002, Textile Museum, Washington, D.C.
[56]See "An Introduction to Textile Terms," 1997, Textile Museum, Washington, D.C.

Communication—not only the words but also the style—is an important issue for collaborators. Education and training shape expectations for communication; they can also factor into receptivity to the vocabulary and styles of others. In a productive architectural process, roles are flexible and the many actors can cross professional boundaries and interact in ways that enable creative things to happen. If an architect knows something about structural engineering, and a structural engineer knows something about architecture, they can perform their specialized roles at a sophisticated level of discourse. For instance, the architect can tell the engineer that a column is oversized and know, without being told, that it could be cut in half. They know enough about each other's jobs to communicate across role boundaries. Thus, mechanisms such as crossover books (books that are intended for non-specialist audiences) can be useful; such books boil down the essence of an area for the intelligent and interested novice. However, these adaptations must not be so diluted that real insights are obscured by superficialities.

It is no secret that scientists and artists have widely differing community standards with regard to language and modes of expression and the types of questions to explore. As noted by Michael Mateas, for example, the scientist seeks abstract and objective knowledge, whereas the artist seeks an immediate perceptual experience for the audience.[57] Accordingly, it can be difficult for them to reach consensus on common problems and topics and to establish common understandings.[58] Yet there are also rapid changes redefining practice that are blurring previously rigid boundaries, as collaborators find ways to accommodate their differences. As noted by a reviewer of this report, successful collaborations involve mutual respect and friendship: Each knows enough about the other's field for meaningful conversation to take place, but respects the other's expertise enough to leave specialized decisions to that collaborator. Shared goals, group dynamics, and psychological maturity are more important than complete coverage of required expertise.

[57]Although scientists and artists may have different motivations, and public appreciation may play a relatively greater role in artists' visibility and income, in both cases professional advancement depends heavily on the judgment of peers.

[58]See Denise Caruso, 2001, "Lead, Follow, Get Out of the Way: Sidestepping the Barriers to Effective Practice of Interdisciplinarity," white paper, Hybrid Vigor, see <http://www.hybridvigor.org>.

# RESOURCES THAT SUPPORT CREATIVE PRACTICES

## SKILLS TRAINING

Work in ITCP not only demands new capabilities from many of its practitioners, but also offers novel avenues for learning these skills. That is to say, IT can be exploited both to help technologists and artists learn skills and methods and gain access to tools, and to motivate and educate others, including young people, who might one day become active in the field. This last point is important, because children naturally possess both the experimentalism and the fascination with computers that drive success in this field. Online vehicles are already supporting distance education, including instruction in new methodologies in general and the use of specific tools. Organizations that produce tools are increasingly turning to the Web as the medium of choice for providing educational material, supporting user-directed learning.

There seem to be more resources offering IT skills training and tools than offering arts education, paralleling what some see as an asymmetry in the motivation of artists and technologists to "cross over" into the other domain. There is a belief that, in general, artists can learn IT faster than technologists can learn art, in part because artists are more motivated to use IT as a way to do exciting and distinguished work (e.g., in computer animation). Technologists generally have little general education in art and tend to see the beauty of finding and solving problems in programming and mathematics as their art; in addition, they are paid well in their chosen profession and have less motivation to learn art or design.[59]

An important resource in the mid- to late-1990s was Open Studio: The Arts Online,[60] a national initiative of the Benton Foundation and the National Endowment for the Arts that provided Internet access and training to artists and non-profit arts organizations. According to promotional materials, Open Studio empowered the arts community to "give the Internet a soul," helping artists and arts organizations gain powerful new opportunities to network, strengthen ties to communities, and build new audiences, while ensuring that the online world is a source of creative excellence and diversity.

Technology plays a role in education at Eyebeam Atelier, where the goal is to expose broad and diverse audiences to new technologies and the media arts while simultaneously establishing and articulating

---

[59]Based on a personal communication from Bill Alschuler, School of Critical Studies, California Institute of the Arts, 2002.

[60]See <http://www.openstudio.org>.

new media as a significant medium of artistic expression.[61] Eyebeam accomplishes this objective through three core outlets: education, exhibition, and an artist-in-residence program. The education programs focus on exposing youths, families, and the general public to new-media art using the atelier method, in which an emphasis is placed on studio-based education augmented by technology, and through one-on-one instruction and mentoring. The new artist-in-residence program connects artists with "technology partners," primarily for-profit firms, that provide the technology needed by the artist. The partners share a common goal of exploring the technology's potential in the process of making art.[62]

Relevant online resources are not focused exclusively on the technology side of the ITCP equation. Practitioners can learn elements of artistry as well. For example, mH2O provides the software and samples (short loops of beats, instruments, and vocals) for anyone to create and record music. It also offers a variety of resources including digitized classes with master musicians. For example, users can select from five lessons (on topics such as the "Doodle System" and the "Ooo Bah System") with Clark Terry, a master of the trumpet and flugelhorn, who teaches form, phrasing, articulation, riffing, and other elements of the blues to a group of students at a high school in Connecticut, a project organized by the Greater Hartford Academy of the Arts.

Advances in IT can enable new modes of learning. For example, Maestro Pinchas Zukerman held a videoconference chamber music demonstration and discussion using Internet2 networks and peer networks, CANARIE, and NYSERNet. Zukerman led the class from Ottawa to a talented young string trio in New York. Audience members at Columbia University as well as observers on the Internet could watch the session in real time. A question-and-answer session was held for both in-person and Internet observers.[63]

## WORK SPACES

Appropriate work spaces are an essential ingredient in creative production.[64] People need a comfortable setting offering access to their tools and collaborators. Most discussions of IT work spaces assume the conventional form factor of computing: a screen, a keyboard, and a mouse. Add in all of the normal peripherals of scanner,

---

[61]See <http://www.eyebeam.org/about/profile.html>.

[62]See <http://www.eyebeam.org/artists/index.html>.

[63]See <http://www.columbia.edu/acis/networks/advanced/zukermaninteractive>. Also see *Cultivating Communities: Dance in the Digital Age*, Internet 2, University of Southern California, October 29, 2002, <http://apps.internet2.edu>.

[64]"It is easier to enhance creativity by changing conditions in the environment than by trying to make people think more creatively." See Mihaly Csikszentmihalyi, 1996, *Creativity: Flow and the Psychology of Discovery and Invention*, Harper Collins, New York, p. 1.

printer, telephone, and so on, and the space suddenly needs a desk and a chair. Then information has to be stored in folders, files, drawers, and shelves. Suddenly the work space is an office. Is the artist/ designer studio of the future really a conventional office?

Contemporary work spaces are in flux. During this time of change (or evolution), ITCP practitioners might be best served by flexible and open designs that allow for new configurations to alter the flow of work and communication. Wired spaces (in which there are distributed communication systems for Internet or broadband access) and wireless spaces (areas set up for pervasive access to wireless communications for access to the Internet and to people or devices within the area) are a new part of this landscape. The re-thinking of design for knowledge sharing, through both physical proximity and electronic communication, is an important part of creating new work processes and has to evolve hand in hand with space planning. How can these processes be facilitated in ways that allow for the flexibility and cross-pollination that are desirable in facilities for research and creative production? How can environmental adaptability, and signaling that colleagues are available, be achieved without the suggestion of a surveillance culture? Does electronic networking really reduce "one person–one computer" isolation? How is it possible to create spatial configurations that reduce isolation and foster or enhance discussion?

In the future, devices will get tinier and interfaces will become more complex. The world will have more buttons to push, more gadgets to carry, and/or more systems embedded in the environment (physical or natural) that provide services without direct human interaction. Or perhaps there will be systems for direct input to or output from human brains, possibly through implanted devices. Simply imagining something with visual or physical form could spark an entire sequence of events to occur in the physical world. The boundaries of the real and the imaginary could become obscured. There might be no need for a physical workplace, at least for utilitarian reasons—although there may be essential social needs that are unfulfilled in a virtual workplace. One might just imagine a workplace, and it would appear just as imagined. Early indicators of such phenomena can be found in experiments with virtual worlds, although virtual- and augmented-reality technologies engage a broader range of senses for inputs and outputs than those accessible to ordinary office or home computing systems.[65]

The desktop will most likely have to change to enable a more sophisticated dialogue with digital media. A variety of technologies offering three-dimensional graphics, voice and touch input and output, rapid macro-fabrication capabilities, and terabytes of storage all point to a potential diversification of tasks involving IT and an in-

---

[65]See Computer Science and Telecommunications Board, National Research Council, 1997, *More Than Screen Deep: Toward Every-Citizen Interfaces to the Nation's Information Infrastructure*, National Academy Press, Washington, D.C.

crease in activities that seem to be more versatile and challenging. In the end, environments have to be engaging on many levels for users to have the necessary impetus to respond with impassioned content—to be creative.

Virtual spaces can be architectures for collaborations that allow multiple users to talk and share work and work space across geographical territories. These capabilities are changing the nature of collaborative work as well as the markets and audiences for it. The ability to access and work with a niche group that is broadly distributed geographically allows for new kinds of practice to evolve. Skills that were previously determined locally no longer need be, and audiences that once had to be concentrated at a local level to make the activity economically viable can now be spread over a wide geographical area. These practices and methods of communication are beginning to generate new tools and work methods as well as new territories of content. Academic institutions and research facilities can become leaders in this area, empowering people to experiment in a non-prescriptive way.

As one example, the entire economy of music production has been transformed by digital technologies. Large commercial studios and studio musicians are vanishing as the home studio becomes the standard for production in both the commercial and non-commercial spheres. These studios can now access a level of technology previously unavailable to the individual and will certainly produce new forms of sound design. But technology and social infrastructures have to be developed carefully to avoid jeopardizing social interactions, in which people learn how to play with each other in groups. Access to tools and to other musicians through electronic networks has tremendous potential. It is possible to think of situations, in academic and research institutions as well as in commercial and non-profit production facilities, in which musicians and composers can collaborate both physically and electronically, or virtually, with enhanced potential for discussion and research. A convention of distributed performance has developed—a concert with some players at one site and some at another, or people waving across videoconferencing systems. But the increasing diffusion of broadband technologies[66] has begun to suggest a more complex and sophisticated set of possibilities for multisite performance, including collaborative production and development and new methods of distribution.

---

[66]See Computer Science and Telecommunications Board, National Research Council, 2002, *Broadband: Bringing Home the Bits*, National Academy Press, Washington, D.C.

# 3 | Advancing Creative Practices Through Information Technology

H ow can computer scientists support new artistic and design practices? How is computer science (CS) research and development being stimulated and altered by emerging practices at the intersection of information technology (IT) and the arts and design? What are the prospects for new research directions that are interesting and useful for both computer science and the arts and design? Answering these questions requires a close look at the relationship between IT and the arts and design. That examination begins with this chapter, which focuses on the design, applications, and implications of the tools of IT from the perspective of the arts and design worlds, and it continues in Chapter 4.

• • • • • • • • • • • • • • • • • • • • • • • • • • • • • • • • • • •

## STRANGE BEDFELLOWS?

One of the more obvious ways in which IT and the arts and design interact is in the use of technology to extend the expressive range of and modes of access to existing genres of the arts and design: Examples include Web-based art and hypertext, opera staging using new sensing and video technologies, musical compositions that feature both newly created instruments and interaction styles, and textile design and production based on digital weaving techniques. Given experimentation to date, it is clear that new tools developed by computer scientists can be immediately applied by artists and other creative practitioners within a wide array of contexts.

But there are further important implications of information technology and creative practices (ITCP) for computer science research and development. Box 3.1 provides context on the nature of that research for those who may be unfamiliar with it. Rather than using computational technology as black boxes for arts and design applica-

---

**BOX 3.1**

**Computing, Computer Science, and Research**

Computing is rooted in the discipline of computer science,[1] which studies information and computational processes—including the representation, implementation, manipulation, and communication of information.[2] There are relatively few inherent natural limitations in computer science, as compared with other science and engineering fields such as physics, chemistry, and mechanical engineering. This means that the definitions—and, by extension, the capabilities—of computing and IT often do not have an obvious finite upper bound for expansion. There are, however, practical constraints on the available capabilities of computer hardware and software. In addition, the growing reach and complexity of computers and networks (such as the Internet) have heightened the risk and impact of system failures and created formidable challenges in areas such as security and the management of intellectual property. Efforts to improve computing capabilities are central to IT research. The key intellectual themes in computer science and engineering are algorithmic thinking (i.e., about rules for processing information), representation of information, and computer programs.[3] Some IT research lays out principles or constraints that apply to all computing and communications systems; other studies focus on specific IT systems, such as user interfaces.

---

[1]The history of computing (which involves other fields such as electrical engineering) has been widely documented; see, for example, Computer Science and Telecommunications Board, National Research Council, 1999, *Funding a Revolution: Government Support for Computing Research*, National Academy Press, Washington, D.C.; and the public television documentary *Triumph of the Nerds*, transcript available at <http://www.pbs.org/nerds>.

[2]These aspects of computer science will be discussed in the forthcoming report of the Committee on the Fundamentals of Computer Science—Challenges and Opportunities, Computer Science and Telecommunications Board, National Research Council.

[3]See Computer Science and Telecommunications Board, National Research Council, 1992, *Computing the Future: A Broader Agenda for Computer Science and Engineering*, Juris Hartmanis and Herbert Lin, eds., National Academy Press, Washington, D.C.

---

tions, some are engaging in IT research as a form of art and design practice itself. This activity is less like historical arts research and more like computer science research, although it asks radically different kinds of questions and introduces a variety of methodologies generally unfamiliar to computer scientists. In certain areas of research, particularly human-computer interaction and artificial intelligence, there may be a convergence developing between new trends within CS research and the work of "outsiders" who bring in fresh perspectives. Unlike the use of computers for particular applications, this intersection of art and design and IT research leads to some deep and fundamental rethinking of CS research and what it is about in the first place. The intended outcomes go beyond making new tools for art and design practice, though that may be one outcome, to arrive at a fundamentally new way to do research—a true hybrid.

Both of these perspectives on interaction are important for the future of ITCP. In practice, they are intermingled. On the one hand, developing tools for new kinds of practices can lead to fundamental insights into the tool-development process. On the other, artists and designers who get their hands dirty in fundamental CS research are able to build new tools and applications that can be useful for the arts and other creative domains.

The implications of this new medium for artistic and design practice run deeper than simple application. In previous communications revolutions, a new medium that was widely adopted not only added new possibilities for artistic and design expression, but also changed the way older media were used.[1] Communication media influence the relationship between sense and bodily skill, and they alter the way in which artists and designers reason or feel about time and space.

If one were to think of such a shift as a radical break, in which a developing medium brings an entirely new art form into existence with little relationship to historical precursors, the naïve response for computer science researchers would be simply to generate as many new-media forms as possible as a way of advancing the opportunities for art and design practice. One technically oriented current in the contemporary avant-garde has indeed sought to push technology forward, in order to discover new expressive possibilities and genres that can be conceived only with the help of advanced technologies. Technological advances exert a strong pull, challenging artists and designers to conceive new expressive forms that can take full advantage of ever-increasing processing speeds and bandwidth rates. Both this push on and pull of new technologies focus on the new possibilities created by those technologies, rather than on the needs and perspectives of art and design practices using "old" (and by implication out-of-date) media.

It is a mistake to overemphasize the entirely new digital worlds that are uniquely possible using computers, as though the adaptation of already existing content or art forms to the new medium were only a lesser, transitional stage on the way toward the more significant discovery of purportedly new, essentially digital art forms.[2] What tends to be overlooked, both by the modernist artist's and designer's technology push and the information technologist's pull for advanced content, are the subtle and by no means trivial processes of change occurring in the traditional art and design forms as they adjust to and begin to find their own responses to information technologies. See Box 3.2 for an overview of how technology has influenced music while traditional activities have persisted.

Precisely because these developments are less visible, whether judged by criteria of radical artistic and design novelty or technical

> The naïve response for computer science researchers would be to generate as many new-media forms as possible for art and design practice.

---

[1]The rise of broadband continues that phenomenon. For example, one recent study found that, because of Internet use, 37 percent of broadband users watched less television, 31 percent spent less time shopping, and 18 percent spent less time reading newspapers; nearly 90 percent said that the Internet had improved their ability to learn. See <http://www.pewinternet.org/reports/toc.asp?Report=63>.

[2]See J.D. Bolter and R. Grusin, 1999, *Remediation*, MIT Press, Cambridge, Mass., pp. 48-50. Note that this situation may differentiate the arts from various practical activities (e.g., office and factory work), where early mechanization was, indeed, a step toward broad reconceptualization of different activities. Tradition and continuity are recognized across the arts.

**BOX 3.2**

**Technology, Music, and the Evolution of Expectations**

Consider the history of music. Early music was not written—there was only the performance. The evolution of the musical score, a standardized interface between the composer and the performer, made it possible for the composer to achieve wider visibility by having his music performed numerous times by different groups. One tradeoff for this wider distribution was limited scope—only music that corresponded to this standard score could be written down (excluding all but Western music). Another was that the performance was still transient. It is said that Bach and Mozart were wonderful performers, but records are preserved mainly in listeners' accounts, and in the occasional attempts by both masters to notate transcripts of their more spontaneous fantasy.

With the advent of recording, the performance itself could be captured. And the relationship between the composer and performer shifted. More improvisational styles such as jazz, blues, and rock can now be captured as a performance, not as a score, and the performer of current popular music is expected to perform his own music, not "cover" the music of others. We are back to the era of Bach or Mozart, but with global scale—the benefits of wider distribution now extend to other forms of music that cannot be written down. In turn, recording technology also enabled a new form of music—music that was created in whole or in part with synthesized, recorded, electronic signals, defying both transcription and performance. Thus, the introduction of recording raised new questions about music as a creative process and the definition of a live performance. There is something incongruous about attempting to trigger all the pleasure and excitement of a live gathering of an audience to look at a computer and a bunch of speakers up on stage. To some extent, of course, this is because we know that there will be no spontaneity in the performance—that creativity has already happened. This is what recording did to the creative aspect of music.

The computer is now doing much more. First, the computer can now be the performer. Digital technology allows composers to craft their own instruments and to create a performance using only the computer. Computer-driven pianos now replay a performance with many of the nuances of the original pianist's keystrokes. Interactive computer programs can participate in the performance, producing and manipulating sounds in response to a performer's actions. Digital technology, through interactivity, helped bring spontaneity back into the performance of electronic music. What seems to be evolving is a remodularization of the creative process, the resolution of which is a creative act in its own right, and one that cannot escape the feedback and interaction with the listener. The role of the listener, the audience, illustrates how advances are gated by a social process.

Secondly, the computer can capture a performance and make it permanent. What the recording industry did for music, the computer can in principle do for a more dynamic and multimedia creation. Consider also the emergence of the disc jockey who does a live remix of pre-existing music (including electronic music): This seems to be a reassertion of the persistent value of the spontaneous process of performance, involving real-time feedback between the performer and the audience. Here, again, questions arise about how to view the "dilemma" of the computer as a part of the creative process.

progress, they receive less attention than more obviously glamorous showcases do. But these subtle developments are no less important for the long-term ecology of digital culture, suggesting limitations of, and different possibilities for, the development of technology as a medium.

For example, tacit knowledge—unformalized and probably unformalizable knowledge such as design methodologies or embodied skills such as drawing or dancing—has always played an important role in the arts. Yet IT has by its nature had severe difficulty codifying the most subtle and refined of such artistic practices. The ineffable human feel can be simulated acceptably through clever tricks, but the danger all too often is that consumers of technology, including artists and designers, will accommodate themselves to the reduced expressive bandwidth afforded in easy-to-use interfaces, as discussed below.[3] The emerging research paradigm for embodied interaction in human-computer interaction (HCI) is one opportunity for a different style of interface perhaps more compatible with highly skilled art practice.[4]

• • • • • • • • • • • • • • • • • • • • • • • • • • • • • • • • • • • • •

## TOOLS NEEDED TO SUPPORT CREATIVE WORK: HARDWARE AND SOFTWARE

Computer and communications hardware and software are the tools of ITCP and the means by which almost all digital media are created and manipulated. These tools can do many things,[5] each and all of which may be embraced in ITCP. A simple list of gross capabilities would include:

• Automation of processes such as drawing, composing, editing, and so on (assorted software);
• Handling, representing, and displaying or performing information (databases, browsers, displays and speakers, printers, projection systems, and so on);
• Analysis of information and phenomena (visualization and sonification, modeling and simulation, artificial intelligence and learning systems);

---

[3]For sociological analysis of creative effects of the commoditization of musical instruments in digital forms, see P. Théberge, 1997, *Any Sound You Can Imagine: Making Music/Consuming Technology*, University Press of New England, Hanover, New Hampshire.

[4]Embodied interaction has been a theme in digital arts at least since Myron Kreuger's VideoPlace of the early 1970s. See P. Dourish, 2001, *Where the Action Is: The Foundations of Embodied Interaction*, MIT Press, Cambridge, Mass.; and P. Dourish, 1999, "Embodied Interaction: Exploring the Foundations of a New Approach to HCI" (see <http://www.ics.uci.edu/~jpd/publications>). An example of a new body-centered interface for art is described in Steven Schkolne, Michael Pruett, and Peter Schröder, 2001, "Surface Drawing: Creating Organic 3D Shapes with the Hand and Tangible Tools," pp. 261-268 in *Proceedings of CHI 2001*, ACM Press, New York.

[5]Broad interpretations include not only components and artifacts but also their theoretical underpinnings and digital content. The Computing Research Repository, for example, lists 34 subject areas; see <http://xxx.lanl.gov/new/cs.html>. The Computing Research Repository is an online archive of computer science research results that uses the Internet to allow access to technical reports, conference papers, and other work on a near-real-time basis.

---

**BOX 3.3**
**One Set of ITCP Technologies**

One perspective on technologies relevant to ITCP work is provided by Carnegie Mellon University's Entertainment Technology Center (ETC), which offers a master's degree jointly conferred by the College of Fine Arts and the School of Computer Science. Entertainment technology "requires a fluid definition, necessitated in large part by advances in technology that are making possible ever-new entertainment experiences and venues. What was meant by the phrase entertainment technology as recently as a year ago requires redefinition in light of recent developments in both technology and entertainment."[1] Nevertheless, the following listing is provided:

- Networked and free-standing interactive computer games
- Avatar creation and utilization
- Massive multiplayer online games
- Digital entertainment
- Specialty venues such as theme parks
- Motion-based rides
- Console and PC interactive game design
- Creation of unique input devices
- Virtual reality utilizing head-mounted displays
- Other forms of virtual reality technology
- Wearable computing for entertainment purposes
- Massive immersive display environments
- Interactive robot animatronics
- Synthetic interview technology
- Speech recognition
- Augmented reality
- Telepresence for entertainment and education purposes
- Digital production and postproduction
- Sound synthesis, surround sound, three-dimensional sound, and streaming audio
- Development of haptic devices (e.g., force feedback)
- Entertainment robotics.

---

[1]See <http://www.etc.cmu.edu/about.html>.

---

- Connection to the physical world (sensors, microphones, digital cameras, actuators, robots, human interfaces, systems for interactivity); and
- Communications (telephony, television, the Internet, and assorted underlying capabilities, from wireless to broadband connectivity).

Ultimately, IT acts on information typically associated with products of the human mind (pictures, music, and ideas), although IT must also address data and information in less processed, intermediate states.[6]

---

[6]This concept is derived from Computer Science and Telecommunications Board, National Research Council, 1992, *Computing the Future: A Broader Agenda for Computer Science and Engineering,* Juris Hartmanis and Herbert Lin, eds., National Academy Press, Washington, D.C.

Human-computer interaction specialist Ben Shneiderman argues that IT for creativity support falls into eight categories: searching, visualizing, consulting, thinking, exploring, composing, reviewing, and disseminating.[7] Because the currency of IT is bits, IT tools can handle any mode or medium, and they can integrate any combination of media—although sophistication has costs in complexity and dollars, and for technical and/or economic reasons, not all desired effects are possible (see "Economic Realities" below). For a concrete illustration of a tool set inspired by entertainment applications, see Box 3.3. Lists like these show the broad range of IT that may be linked to some aspect of the arts or design through ITCP; no list conveys all the creative possibilities inherent in the use of IT.

By definition, tools are supposed to be helpful, but like other tools, IT tools have shortcomings. The insights into the nature of IT that collectively constitute fluency, as described in Chapter 2, include an understanding of the limitations of the tools and the constrained nature of the typical software design. The pervasive hype about IT in the mass media may, by contrast, feed unrealistic expectations. Although artists and designers share frustrations with other user groups, their perspectives, like those of other users, may help illuminate new paths for IT research and development. This situation was recognized in a recent special issue of a leading computer science journal:

> Many significant advances in research on human creativity have occurred, yet today's tools often contain interface elements that stymie creative efforts. A discontinuity exists between technology tools and our ability to interact with them in natural, beneficial, and most importantly, for this discussion, creative ways.[8]

Tools vary in terms of the computing and programming skills required to use them. As long as the tools required to produce computer-mediated work are programming tools, the result will be programmer-created design. That is not a bad thing—and in some cases the result(s) can be wonderful. But it does mean that an investment must be made in learning, which is somewhat like the requirement to master other tools used by artists, but also different, because of factors such as the range of features and capabilities available from software and the relatively rapid change in technology. There is a great distance from the paintbrush or piano to programming in C++. As emphasized in Chapter 2, fluency can provide a middle ground between simple acceptance of a tool and expertise in programming.[9] Although the

> There is a great distance from the paintbrush or piano to programming in C++.

---

[7]See Ben Shneiderman, 2002, "Creativity Support Tools," *Communications of the ACM* 45(10): 116-120.

[8]Winslow Burleson and Ted Selker, 2002, "Introduction (Special Issue: Creativity and Interface)," *Communications of the ACM* 45(10): 88-90.

[9]Of course, new approaches to programming that simplify it may be helpful—assuming that the results do not present the same concerns that software packages do about built-in constraints.

discussion in this chapter focuses on software and to a lesser extent hardware for creating artifacts and performances, it should be noted that other tools consist of content (images, sounds, text, and so on) repositories. There the concerns center on the accessibility of the content—involving indexing, permission to use (or ease of obtaining same), and so on. Yet still other tools support public access.[10]

## HARDWARE AND SOFTWARE TOOLS: A MIXED BLESSING

Information technology has obviously proved useful and accessible enough to give rise to ITCP in many guises. In the process, a number of observations have emerged about the nature of IT and the adaptations that artists and designers make in using it. The concerns gleaned from artists and designers help to explain why it takes a long time to integrate a new technology into the making of non-trivial art or design work.

Because all computers are universal machines, more advanced designs can replace simpler ones, providing improved performance without changing the basic functionality of computation. Software is even more mutable in that it has almost no physical constraints. Because of rapid changes in hardware capabilities, software rarely matures to a stable configuration. On the positive side, this pace of change provides frequent opportunities to incorporate improved support for creativity in new systems. But for most users, especially in relatively resource-poor areas such as the arts and humanities, this constantly changing tool set is difficult to master. For well over a decade, it has been the lament of artists and designers that the intellectual and financial demands of constantly updating tools and playing technological catch-up results in low-quality work and burnout.[11] Smart artists, observed a reviewer, resign from the Moore's law rat race. These conditions suggest that it is reasonable to expect a wide range of willingness and ability among artists and designers to retrain and to upgrade their tools; how this will affect ITCP remains to be seen.

Developers of software tools that can support creative practices have a number of variables to consider, all of which may affect the

> It takes a long time to integrate a new technology into the making of non-trivial art or design work.

---

[10]See tool characteristics in Sharon L. Greene, 2002, "Characteristics of Applications That Support Creativity," *Communications of the ACM* 45(10): 100-104.

[11]An anecdote shared by a reviewer featured a senior colleague who reported that when he discovered the Amiga computer, he felt sure he had found the tool with which he would make his magnum opus. After several years of learning about and constantly upgrading his Amiga tools, he had become an Amiga expert but as yet had produced no work. Much to his chagrin, the machine then became obsolete. More generally, in the mid-1990s a committee of media arts faculty prepared a document, colloquially referred to as the "'burnout" document, that outlined the new and unrecognized loads on media arts faculty. It was endorsed by the College Art Association, Inter-Society for the Electronic Arts, and Special Interest Group on Computer Graphics and Interactive Techniques.

ways in which users interact with the tools—and through the tools, their own work—and the ease with which users can produce original or even groundbreaking works with those tools. To name just a few, tool developers dictate the user conceptual model (or metaphors) exposed by their tools, the amount of structure supported or imposed on the work process and product, the number and kinds of different presentations and representations of content supported, the kinds of manipulations directly implemented, the openness and extensibility of the tool at various levels, and the levels of abstraction afforded. The developers' decisions often reflect attempts to make the system easy to use,[12] or simplifying assumptions about what users want, but they can also greatly affect what users produce.[13] Very little research has explored the relationships between these design decisions and the fitness of the tool for various kinds of ITCP work—what follows are some observations made by the committee.

In an effort to get work done, it is only natural to follow the path of least resistance established by the tools that are available, so these tools play a very important part in how users conceptualize their work and assess their options. The tools may also leave traces: Architects may look at a building and detect which design tools were used, and musicians may hear new pieces and detect which composition tools were used.[14] Thus, IT may act as a flywheel in at least some contexts of its use. Artists with IT fluency recognize these practical realities and their roots in the values and procedures of computer science; they work with or around the limitations of the tools.

In the popular style known as object-oriented programming, a programmer creates a new kind of object (concretely, a data structure that describes something) and defines the operations that other programmers can perform on it. Other programmers that use the object cannot get at the data structure itself, but only at the operations that the object-creator defined. This gives the object-creator the ability to come back later and change the actual data structure (perhaps to make it more efficient) without having to coordinate with any other pro-

> Tool developers dictate the user conceptual model (or metaphors) exposed by their tools.

---

[12]An early illustration in a number of (non-artistic) fields was statistics; statistical software (and later spreadsheets) popularized and helped to disseminate a variety of quantitative methods used in different applications. More recently, software packages have begun to implement neural networks, face recognition, data acquisition, or image display. Because these functions can be technically difficult to create, careful packaging can allow others to use advanced technology with less effort.

[13]For example, the literary theorist has written, with reference to hypertext authoring environments: "The strength of metaprograms is that they take away most of the pain involved in programming an application from scratch; . . . [their] weakness is that they limit the programmer by presenting a predefined range of operations that the programmer must use. . . . This may be compared to pre-modern modes of authorship, in which the author could use predefined paradigms to produce a genre text, without much creative effort." See Espen J. Aarseth, 1997, *Cybertext: Perspectives on Ergodic Literature*, Johns Hopkins University Press, Baltimore, Maryland, p. 173.

[14]In other contexts, such as education, people have noted how the presentation software PowerPoint shapes and constrains their choice of content and organization for public speaking.

grammers. The drawback is that the user can perform only the operations the object-creator thought of. If the user wants to add a new operation, he or she cannot, if doing so requires getting at the base data representation. It is much easier to design a tool for a well-defined task than for non-deterministic, creative work, which cannot be reduced to a set of tasks.

As one example of how these design decisions can affect the utility of tools, consider an image-editing program that provides a "blur" feature with a capability for "more" and "less" control. Technically, this is a simple mathematical convolution on the image, but the tool designers chose not to expose the numeric parameters for this operation. The benefit is that an operation assumed to be common—blurring an image—is simplified. The tradeoff is that other convolutions—which might prove interesting or useful to a user—are impossible without some other software mechanism.

One design point that provides good support for explorations of ITCP is to use a small number of concepts in a general and flexible way combined with openness and interoperability with other tools. A good example is the interactive music language "Max," a full visual programming environment that has developed a community with at least three ranks of creative users. The least technical of these would be a performer who develops skills as an interactor with a particular set of patches (configurations), much like a musical performer would practice her instrument. Users may then learn to make their own programs, by patching together already existing objects in unique ways. Such patches then become instruments that permit a high degree of virtuosity for use within particular genres (like techno-music or interactive dance).

Creating a good tool requires an understanding of the problem area and often experience at the cutting edges of artistic and design and technical disciplines, just to understand the underlying concepts and representations that the tool will address. Tool making also can benefit from combining technical expertise in designing and implementing software with a healthy dose of common sense to arrive at something general enough to be useful but simple enough to be used.[15] Problems sometimes arise when tool design assumes too much separation of idea from expression.[16]

> It is much easier
>
> to design a tool
>
> for a well-
>
> defined task than
>
> for non-
>
> deterministic,
>
> creative work.

---

[15]People familiar with computer science might consider Donald Knuth as an early explorer of ITCP. Venerated for his understanding of software, Knuth has done work on fonts and electronic composition that is recognized broadly.

[16]Wright points out that the separation between idea and expression in software—which can degenerate to the software supplying all the ideas—is one example of a more general problem with creative software tools. These tools most commonly "aim to mirror internal creative process by organizing it into an external data process or structure. The software is a system of menu commands and options which seeks [is designed] to match an internal model of creativity as a process of decision making that seeks to approximate an ideal artistic goal." But artists and designers "do not know in advance what they want before they start—the creative process is actually a process of

A tool may offer just the right functionality, but not at the right component level. For example, an image can be manipulated through the use of a comprehensive interactive application such as PhotoShop, or by typing a command to run a specific program such as one of the Unix ppm utilities,[17] or by calling a function in an image processing library. A software library is of little use to a non-programmer, whereas a full-blown application is of little use in creating a custom software program. Sometimes, a full-blown application can be made to perform like a custom software program through scripting, and sometimes software libraries or other components can be integrated with an application via a plug-in capability (see below). More could be accomplished with tools designed to suit a range of different tasks. Ensuring this flexibility calls for either a variety of tools or a hierarchical structure that allows access to tools at various levels of abstraction, permitting a migration path from intelligent enhancement at the novice level to customization for experts (as illustrated by the language "Max" described above).

Because creativity is associated with novelty, comprehensive tools for creative work will be neither possible nor necessary to develop, any more than it is necessary for a pencil to include all functions for drawing. It will always be the task of the innovator to create new tools from components, to create new applications, and to create new artifacts by using tools in unanticipated ways. In this respect, ITCP will draw from both artistic and design and computer science traditions with experimentation. Small professional communities that share experiences and talent often develop effective tools—this phenomenon is evident in the growth of computational science and the development of the Web by and for physicists, and it has been observed in humanities fields as well.[18] Many small tools are developed by individuals or small teams that are able to keep their tools focused and coherent.

Design choices related to tool extensibility may be particularly important for broadening participation in ITCP and tool development. Programming language extensions (also called scripting languages) allow end users to customize applications. For example, computer music and animation programs often use text-based scripts or scores to describe music and images. Users can create scripts either by editing text directly or by writing programs to generate scripts. Another way to customize tools is with a plug-in, which is a software compo-

> Comprehensive
>
> tools for creative
>
> work will be
>
> neither possible
>
> nor necessary.

---

playing and 'visual thinking' that leads to a variety of interim 'solutions' and modifications of the original 'problem' . . . ." See Richard Wright, 1999, "Programming with a Paintbrush: A Study in the Production Culture of the Moving Image," *Filmwaves*, Issue 12.

[17]These programs perform a wide range of image-processing tasks including format conversion, scaling, filtering, and color adjustment.

[18]See American Council of Learned Societies, 1998, *Computing and the Humanities: Summary of a Roundtable Meeting*, Occasional Paper No. 41, available online at <http://www.acls.org/op41-toc.htm>.

nent that extends an application by implementing a new function—a much simpler approach than writing an entirely new application. Plug-ins are common for image- and audio-processing programs; they exist for such popular design tools as PhotoShop and Auto-CAD.[19] Although commercial image and video editors may not support functions desired by researchers, artists, and designers working at the creative frontiers, a plug-in architecture enables users to extend editors with new custom functions. Note that this capability presents its own challenges to implement: A program like PhotoShop that allows other programmers to create plug-ins has to reveal a lot of its internal structure. This allows more third-party creativity (by those who have the programming skills), but it constrains the maintainers of PhotoShop to preserve all this internal structure as they move from revision to revision. Balancing these tensions is an act of creativity in applied computer science. Generalizing from the discussion of plug-ins, creativity can be supported by systems that are extensible, a feature that is often associated with a modular design (a plug-in is a module), to support a broad range of users and their tasks. Modular designs require software "hooks" to which new capabilities can be added. Recently, however, as programs have grown to help users accomplish more tasks with less expertise, they have tended to hide their inner workings and to inhibit flexibility to do things that were unanticipated by the designers of the program.

Open application program interfaces, scripting languages, and plug-in architectures are no panacea, however. Decisions about what to expose (the level of abstraction, the set of functions, access to data, and so on) will still influence the ways in which users can extend a tool. Also, taking advantage of these forms of extensibility requires expertise in programming. Research could illuminate alternative approaches, exploring how to make extensibility and alternative configurations more cost-effective and how to implement virtual tuning knobs that allow artists and other users to make various adjustments.

In the hardware area, sensors[20] often require significant engineering expertise and are difficult to interface with popular operating systems. Ready-to-use designs, circuit boards, and specialized operating systems can simplify the use of hardware sensors in many interesting applications,[21] and there is evidence that at least elements of a

---

[19]An example is the RealOne Player (by RealNetworks) for downloading and viewing video segments on the Internet. See <http://www.real.com>. Other examples include plug-ins that allow browsers to read proprietary files, such as those in Portable Document Format (PDF).

[20]Sensors can be applied in many contexts relevant to ITCP work, as described in subsequent sections of this chapter.

[21]See, for example, information on prefabricated robot microcontrollers, such as the Handyboard, at <http://www.handyboard.com/>.

relevant tool base are emerging in IT research circles that may be valuable for ITCP.[22]

Inasmuch as artists' needs depart from commercial offerings, or inasmuch as specialized tools are kept proprietary, as with some instances in the competitive animation arena, artists' work in ITCP may require support for experimentation in tool development. New software tools could be developed that would enable users to build their own software tools—collaboration between artists and designers and computer scientists could aim at a meta-toolkit that would offer ease of use plus flexible, extensible results, and an Internet-accessible repository of available software and hardware tools as well as guidance on what it takes to use them would broaden access and experimentation.[23] Support for programming by people lacking full-fledged programming skills could be part of this kit.[24] For effective development to take place, it is important to understand what shape the desired ITCP software will take—what interfaces it should provide, what functions it should support, and so on. Hence, building research prototypes would be a reasonable first step; such prototypes are going to be built by motivated practitioners or researchers rather than by commercial software developers—who may, of course, transform attractive prototypes into products.

As with other instances of user-generated IT tools or the broader phenomenon of popular culture finding new purposes for tools,[25] ITCP tool-development efforts may prove to have broader appeal. The rationale for encouraging this activity echoes the rationale for encouraging research on usability: Steps designed to benefit a small constituency (e.g., people with various disabilities) may prove to benefit users in general.[26] Research oriented to enhancing creativity is high-leverage.

> Collaboration between artists and designers and computer scientists could aim at a meta-toolkit that would offer ease of use plus flexible, extensible results.

---

[22]See, for example, the TinyOS work based at the University of California at Berkeley. It aims at developing an open experimental software/hardware platform for network embedded systems technology research that will dramatically accelerate the development of algorithms and services and their composition into challenging applications. Small, networked sensor/effector nodes are developed to ground algorithmic work in the reality of operating with numerous, highly constrained devices. See <http://webs.cs.berkeley.edu/tos/>.

[23]Rhizome.org, for example, has incubated an open-source software archive for artists (artists can submit open-source code to an online database, download code, discuss it, and so on).

[24]There may be capabilities to borrow from the education community, such as Logo and its derivatives, or from efforts aimed at broadening use of the Web. Or there may be insights to glean from other kinds of tools. For example, Mathematica is a program to perform mathematics. An important component of Mathematica is its programming language, which allows tedious tasks to be automated, presumably freeing up some of the user's time for more exotic pursuits. See <http://www.mathematica.com>.

[25]As William Gibson puts it in *Neuromancer* (Ace Books, New York, 1984), the street finds its own uses for things.

[26]Computer Science and Telecommunications Board, National Research Council, 1997, *More Than Screen Deep: Toward Every-Citizen Interfaces to the Nation's Information Infrastructure*, National Academy Press, Washington, D.C.

---

**BOX 3.4**

**Sketchy Interfaces**

"Sketchy" interfaces provide the ability to describe objects with ambiguous types, sizes, shapes, and positions quickly. The ambiguity encourages a focus on larger concepts, rather than details such as font size or precise alignment. James Landay has developed a tool called "Sketching Interfaces Like Krazy" (SILK) that allows designers to create prototypes of interface designs by sketching. The difficulty with previous interface tools is that they have required designers to specify color, shape, size, and orientation of the elements of an interface precisely from the start. This requirement focuses the designer's and his or her test users' attention on aspects of the design that are irrelevant in the early stages of design. As a consequence, many designers prefer old-fashioned paper sketches. Based on a survey of interface design practices, SILK attempts to combine the positive imprecision of paper sketches with the benefits of storage, searchability, and interactivity that information technology can provide. Designers can roughly sketch and storyboard their ideas on a sketch pad, while SILK automatically makes the sketches interactive, allowing them to be used for testing.[1]

---

[1]For further discussion on SILK, see <http://www.cs.berkeley.edu/~landay/research/publications/CHI96/video.html> or James A. Landay, 1996, *Interactive Sketching for the Early Stages of User Interface Design*, Ph.D. Thesis, Carnegie Mellon University, Computer Science Department, available online at <http://reports-archive.adm.cs.cmu.edu/anon/1996/abstracts/96-201.html>. Also see <http://www.computer.org/computer/homepage/march/cov_feat/0301silk_side.htm>; James A. Landay and Brad A. Myers, 2001, "Sketching Interfaces: Toward More Human Interface Design," *IEEE Computer* 34(3): 56-64; and James A. Landay and Brad A. Myers, 1995, "Interactive Sketching for the Early Stages of User Interface Design," pp. 43-50 in *Proceedings of CHI '95*, Denver, Colo., ACM Press, New York.

## SUPPORT FOR FLEXIBILITY, EXPERIMENTATION, AND PLAY

IT facilitates the rapid design and creation of digital artifacts and their equally fast refinement.[27] Creative design is often an iterative process in which many ideas are considered. Sometimes a sketchpad is much more effective than a computer-aided design (CAD) tool because the precision and detailed specification required for CAD are not needed in the early stages of design. The computing literature gives a number of examples of "sketchy interfaces" that allow designers to make rough prototypes to explore a design space rapidly. There is also evidence that "sketchy" designs tend to elicit more higher-level comments ("Why do you need this function?" as opposed to "I don't like the color scheme"). These and other examples illustrate how tool designers vary the kinds of structures imposed on the work process and product in their tools, often to facilitate experimentation, improvisation, and flexibility. See Box 3.4.

---

[27]"Support pain-free exploration and experimentation (a 'sandbox' mode). There should be an easy way to undo and redo all or part of one's work. There should be no big penalties for mistakes, and there should be meaningful rewards for success. There should be immediate and useful feedback for one's actions, promoting a sense of control." See Ben Shneiderman, 2002, "Creativity Support Tools," *Communications of the ACM* 45(10): 102.

In the right hands, the right computer tools can enable improvisation, which is an important expression of creativity in response to a dynamic situation. Just as a jazz musician improvises by listening and responding to a dynamic musical context, all artists improvise with media in creative ways; on the computer science side, the synthetic nature of software implies that improvisation played a role in its development. At least for some artists and designers—and for many computer scientists and other scientists and engineers—computers carry out computations to explore "what if" scenarios. Such computer-based experimentation inspires new ideas, driving the interplay among design, simulation, surprise, and creation. Tools support improvisation when they offer interactive design, revision, and elaboration of partial specifications. Doing this successfully requires careful consideration of the tradeoffs between imposed structure and clean-sheet approaches, among other design parameters. A highly structured musical tool with a rich model of tonal music may be excellent for certain kinds of improvisation and experimentation, but not for others, for example. See Box 3.5.

## THE INTERNET AND THE WEB

The Internet[28] is particularly useful in ITCP work because of several unique features that set it apart from traditional communications systems, such as the public switched telephone network and the cable and broadcast television systems.[29] The Internet's design encourages innovation at the edges by users, allowing a relatively unrestricted set of applications to run over it. By contrast, traditional networks are centrally developed and managed and historically have limited what users can do with them. Connection, interconnection, and innovation in facilities and services are relatively easy with the Internet, making it possible to use the underlying communications infrastructure more efficiently and inexpensively. These factors have generated a pattern of innovation in Internet technologies and uses associated with a culture of cumulative knowledge-building. Thus, not only are personal computers and larger computer systems attached to the Internet, but so also are televisions, telephones, personal digital assistants, and

---

[28]The Internet has both computing and communications components. The networks that constitute the Internet are composed of communications links, which carry data from one point to another, and routers, which direct the flow of communications between links and, ultimately, from senders to receivers. Routers are computer devices located throughout the Internet that transfer information from a source to a destination. And of course, users access the Internet from a computing device, which may be in a form other than a conventional computer.

[29]For the history of the Internet, see the home page of the Internet Society at <http://www.isoc.org>. See also Katie Hafner and Matthew Lyon, 1996, *Where Wizards Stay Up Late,* Simon and Schuster, New York; and Computer Science and Telecommunications Board, National Research Council, 1999, *Funding a Revolution: Government Support for Computing Research*, National Academy Press, Washington, D.C.

## BOX 3.5
### Improvisatory Interactivity in Music

In his briefing to the committee, George Lewis described his work with improvisational systems.[1] Improvisation characterizes much of human activity, but it has not been explored nearly as much as related activities such as engineering, design, and even creativity generally. At its most basic level, improvisation is a series of reactions to situations, often intuitive, expressive, and spontaneous. Improvisation is about finding structure, not imposing it. In spite of the informal and unplanned nature of improvisation, it can be conducted with consummate artistry, and great improvisers develop their skills over a lifetime of work. Improvisation is an interesting metaphor for human-computer interaction.

Current technology, as practiced in computer music systems, has a long way to go before it will be considered highly skilled, intelligent, or creative. In fact, it can be argued that present systems are successful because of the skill of human performers. Humans can form plans at many time scales, from milliseconds to minutes, and humans are uniquely equipped to listen to and evaluate music as it is created. In contrast, machine musicianship is generally weak and at best skillful in a very narrow range of situations. However, there are interesting and humanly impossible tasks that computers can perform, creating some very interesting musical possibilities.

What does it mean, technologically, to build an improvisation system? First, any interactive system must have sensors and analysis functions. Second, the system must have some method of making high-level musical decisions and generating music. Third, the system must have performance skills, that is, the ability to transform musical information into real-time commands controlling pitch, amplitude, vibrato, and other expressive parameters of sound. Finally, the system must generate sound from this control information, also in real time.

Sensing can be accomplished using a microphone to "listen" to the human performer. Signal processing can detect pitch, amplitude, brightness, and other features, although this is currently restricted to monophonic instruments, those that produce only one tone at any given time. Sensing can also use non-acoustic means, such as optical or electrical position sensors in electronic keyboards, accelerometers, optical sensors, and even a standard computer keyboard and mouse. Once low-level sensor data is acquired, the system must interpret the data in musical terms.[2] The system might parse the data into phrases, build histograms of pitch (indicators of keys and modulation), estimate tempo,[3] or use statistical classifiers to detect emotion,[4] style,[5] or other musical information.

Once processed into some abstract representation of music, sensor data become an input to the improvisation process. The system performs decision making and planning at this stage based on input from sensors, memory of past events, and built-in knowledge.[6] For example, George Lewis's Voyager program is capable of generating music without any input or interaction, but when it senses certain trends in the human improvisation, it can modify its own performance to either go along with, or contrast with, the human musician. This aspect of the improvisation system often draws on formal theories of music and computer science. For example, the system might use Markov models to generate chord progressions, or 12-tone serial techniques to generate melodies. Rule systems are often used to react to sensor data, generating decisions and plans that in turn modify the music generation process.

As high-level representations of music are generated, the music must be performed in real time. Performance can include such things as voicing and orchestration, that is, deciding which synthetic instruments will perform which notes of the music, expressive modulation of various control parameters, and at the lowest level, scheduling accurately timed parameter updates to the music synthesizer. Some recent advances in the understanding of emotion in music, for example, allow a stream of music events to be transformed in ways that express anger, calm, happiness, or sadness.[7]

The final stage is the actual generation of sound. Early improvisation systems often relied on off-the-shelf synthesizers to generate sound from low-bandwidth musical instrument digital interface (MIDI)[8] control information. Now that laptop computers can generate rich sounds in real time, many computer musicians have been attracted to the possibilities of direct control over sound at the signal-processing level. Systems often transform acoustic input from the human performer, giving the improviser a much more intimate and direct connection to the generated sound. In this case, the improvisation might focus more on

FIGURE 3.5.1 Andrew Schloss improvises with an interactive computer music system, using a radio drum and foot pedals. Image courtesy of the Banff Centre.

the character of the sound (timbre) rather than on abstract music structures of pitch and rhythm (Figure 3.5.1).

Many interesting problems arise in interactive improvisation systems. How do improvisers build on local, almost reflexive decisions to create masterful form at a more global level? Can this process be modeled in the computer to create a musical companion?[9] Human perception of music includes sophisticated recognition of patterns and motives. Can systems be extended with better models of perception? Improvisers study and learn. Can computer systems also learn to improvise? Whether discussing biological or silicon systems, computation must take place at many time scales. A jazz drummer produces events with very high timing resolution relative to the speed of high-level cognitive processing. Similarly, computer systems must simultaneously deliver low-latency, hard, real-time signal processing while simultaneously performing high-level decisions at a more relaxed time scale.[10]

Current systems have not solved all of these problems, and they are generally weak at forming plans, performing high-level perception and recognition tasks, and offering sophisticated composition skills. Nevertheless, interactive computer music systems incorporating improvisation have established a new genre of contemporary music. These systems offer a model for human-computer interaction driven by shared goals, high-level task-oriented communication, and creative, situated decision making by both human and computer.

---

[1]See George Lewis, 2000, "Too Many Notes: Computers, Complexity and Culture in Voyager," *Leonardo Music Journal* 10; George Lewis, 1999, "Interacting with Latter-day Musical Automata," *Contemporary Music Review* 18 (Part 3): 99-112 (with accompanying CD); Kristin Palm, 2001, "Making a Point" (review of interdisciplinary, multiartist project at the Point Loma Wastewater Treatment Plant in San Diego, including a site-specific interactive videosonic installation by George Lewis), *Metropolis* (August/September); Guy Garnett, 2001, "The Aesthetics of Interactive Computer Music," *Computer Music Journal* 25(1); Todd Winkler, 1998, *Composing Interactive Music: Techniques and Ideas Using Max*, MIT Press, Cambridge, Mass.; Ben Ratliff, 1997, "Improvisers Meet the Machines" (Review of "Voyager," CD recording featuring the computer music of George Lewis), *New York Times*, October 14; and Joel Chadabe, 1997, *Electric Sound: The Past and Promise of Electronic Music*, Prentice-Hall, Upper Saddle River, N.J.

[2]See Robert Rowe, 2001, *Machine Musicianship*, MIT Press, Cambridge, Mass.

[3]See Masataka Goto, 2001, "An Audio-based Real-time Beat Tracking System for Music with or without Drum-sounds," *Journal of New Music Research* 30(2): 159-171.

[4]See A. Friberg, E. Schoonderwaldt, P.N. Juslin, and R. Bresin, 2002, "Automatic Real-time Extraction of Musical Expression," pp. 365-367 in *Proceedings of the International Computer Music Conference—ICMC 2002*, International Computer Music Association, San Francisco, Calif.

[5]See Roger B. Dannenberg, Belinda Thom, and D. Watson, 1997, "A Machine Learning Approach to Style Recognition," in *International Computer Music Conference*, International Computer Music Association, available online at <http://www.cs.cmu.edu/~rbd/bib-styleclass.html#icmc97>.

[6]See Todd Winkler, 2001, *Composing Interactive Music*, MIT Press, Cambridge, Mass.

[7]See R. Bresin and A. Friberg, 2000, "Emotional Coloring of Computer-controlled Music Performances," *Computer Music Journal* 24(4): 44-63.

[8]See Joseph Rothstein, 1995, *MIDI: A Comprehensive Introduction*, 2nd Ed., A-R Editions, Madison, Wisc.

[9]See Belinda Thom, 2001, *A Customized, Interactive Melodic Improvisation Companion*, Ph.D. Thesis, Carnegie Mellon University, Computer Science Department, Technical Report CMU-CS-01-138, Pittsburgh, Pa.

[10]See Roger B. Dannenberg and Patrick van de Lageweg, 2001, "A System Supporting Flexible Distributed Real-Time Music Processing," pp. 267-270 in *Proceedings of the 2001 International Computer Music Conference*, International Computer Music Association, San Francisco, Calif.

other devices; the future is likely to see many other devices emerge (e.g., music, medical, and kitchen appliances) that will be directly connected to the Internet, opening up many intriguing possibilities for ITCP work.[30] Of course, traditional media, such as television, are also expected to evolve, partly in response to the Internet and partly on their own.[31] Because the Internet is a transnational system, albeit with distributed management, it offers the potential of a worldwide forum. Global digital telecommunication establishes new ways of aggregating expertise and accumulating creative capital, thereby allowing people with similar interests to communicate with each other and share resources.

The World Wide Web, which served to popularize the Internet beginning in the 1990s, has an obvious attraction: It allows media to be distributed at low cost without any special organizational support, and the content can be viewed from anywhere in the world. The Web is quite flexible, offering the capability to publish text, images, sound, and video and to organize the presentation of material in creative ways. In the literary arts, for example, the combination of increasingly powerful desktop computers and the development of the Web has created unparalleled opportunities for people to engage in collaborative work at sites where, for example, hundreds of novice writers may contribute to a narrative. Such opportunities have led to an increasing amount of creative literature on the Web. That said, there are fundamental aspects of the design of the Web that constrain creative work. For example, the client/server model limits interactions either to following links or to performing computations at the server. In addition, the Web browser's distinctive interface containing back and forward buttons works well for some types of work, but not for others, such as work trying to convey complex ideas—another example of the importance of the tradeoffs related to imposed structure. Given that a certain kind of creativity is inspired when faced with limitations, it may be that the limitations of the format can provide a framework for creative work, much as the traditional form of the classical symphony did for classical and romantic music, acting as a structure to both work within and rebel against.

Other models of Web programs suitable for ITCP work are possible. It is not that difficult to program networked applications that

---

[30]This paragraph is drawn primarily from Computer Science and Telecommunications Board, National Research Council, 2001, *The Internet's Coming of Age*, National Academy Press, Washington, D.C.

[31]For example, new roles for public broadcasting (and for broadcasting more generally) may become possible. Television stations in the United States are required by law to convert to digital transmission and program production by December 2006. Initially, it may well be the case that programs are simply converted to digital format in toto. However, the creative integration of digital technology and content may enable many new possibilities for ITCP work. See Lawrence K. Grossman and Newton N. Minow, 2001, *A Digital Gift to the Nation*, Century Foundation Press, New York, available online at <http://www.digitalpromise.org>.

operate away from the biases and constraints of back and forward buttons and that are not subject to the biases and constraints of Hypertext Markup Language (HTML) and Flash (a multimedia plug-in). More computational power is generally available locally, so highly interactive systems such as computer games run on a personal computer. Multiuser games can combine communication over the Internet with local computation. One could even imagine hybrid systems using cable TV, satellite, or telephone communication channels combined with the Internet and local computation. The development of grid infrastructures and applications[32] provides additional capabilities for distributing and sharing capacity and activity, as do two-way protocols like those sponsored by the Universal Description, Discovery, and Integration (UDDI) project.[33] The client/server model is evolving into a many-to-many, dynamic infrastructure.

Recent experience with Napster and other peer-to-peer systems has motivated experimentation among researchers and other creative communities in uses of the Internet for producing and distributing work. Peer-to-peer computing holds enormous promise for leveraging the intellectual and cultural resources of millions of people by allowing connections to be developed independent of centralized servers.[34] Significant proposals include those from scientific researchers, such as Thomas Ray working in the field of artificial life, for research projects to run in the background on computers that are otherwise underused, for example, during the nighttime hours at most large organizations. Notwithstanding constraints on Napster, projects that do not involve copyright issues can still take advantage of peer-to-peer software to carry out research, share information, and share computing resources.[35] Technologies such as peer-to-peer networking also can challenge basic notions of exhibition or participation.[36] Multisite performances in games like Everquest, with persistent virtual realities that endure over days and months, allow participants to create complex virtual societies and emergent dramas. How might conventional museum exhibitions incorporate (or complement or compete with) such new applications of IT?

> Technologies such as peer-to-peer networking can challenge basic notions of exhibition or participation.

---

[32]A grid infrastructure is hardware and software infrastructure that supports wide-scale distributed computing, which enables high-performance applications.

[33]The project creates a platform-independent, open framework for describing services, discovering businesses, and integrating business services using the Internet, as well as an operational registry, according to the project Web site at <http://www.uddi.org>.

[34]Peer-to-peer networking as a technology is not so new per se. The growth of the number of Internet users across diverse segments of society provides the basis for new applications for ITCP work.

[35]*Peer-to-Peer: Harnessing the Power of Disruptive Technologies* (O'Reilly, Beijing and Cambridge, Mass., 2001), edited by Andy Oram, gives a good overview of these possibilities.

[36]Other efforts that may benefit from peer-to-peer computing include scholarly research to which amateurs have made significant contributions, for example in wide-ranging areas such as meteoritics (the study of meteors and mapping of the strewn fields of meteorites), Civil War history, and biography.

The question is how creative practice can really leverage the capability of the Internet—going beyond pretty pictures on Web sites to address the real issues and opportunities that lie on the other side of the (fire)wall. The Listening Post project (which characterizes Internet traffic using sound) described in Chapter 2 represents one of the few ITCP initiatives that delve into the IT infrastructure. As Napster, SETI@Home, ICQ,[37] and countless multiplayer games have proven, people—even millions of people—will take 5 or 10 minutes to download an application if it is sufficiently compelling. There are many opportunities for further work here.

But the commercialization of the Web and associated business and product developments place pressures on Web technology. For example, a break with earlier versions of a software package means that the authors of creative works must spend literally thousands of hours retrofitting those works when a few changes would have made it possible to avoid this tedious labor. This situation is similar to others where backwards compatibility has not been maintained; the larger issue of how vendors choose what features to add or drop reflects what economists refer to as market power, which influences the degree to which consumers have an effective voice. Although other groups of users would also have been inconvenienced by disruptive product changes, that artists and designers complain is a reflection of their comparative lack of financial and technical-support resources to help in adapting to change. A contrasting example of best practices in this context is Rob Kendall's development of HTML authoring software, done in close cooperation with the people who are most likely to use it.[38] See the section "Standards" below for a discussion of standards setting.

Given their roles as content-generators, artists and designers have a special interest in the Internet as a vehicle for content. The Web offers ready access to stored documents, access to which has often required a physical presence (e.g., to view tax rolls, government archives, materials in libraries, and the like), but these materials can increasingly be accessed easily by any Web publisher.[39] The ability of virtually anyone to publish on the Web enables producers of ITCP work to contact audiences more directly. User-provided content plays an important role in the computing and Internet culture, thus offering a unique

---

[37]ICQ (pronounced as "I seek you") is an online instant messaging program developed by Mirabilis Ltd. that is similar to AOL Time Warner's Buddy List and Instant Messenger programs.

[38]See the Word Circuits Connection Muse, available online at <http://www.wordcircuits.com/connect/index.html>.

[39]The electronic documents on the Web are formatted in special languages (e.g., HTML) that accommodate multiple media (text, graphics, audio, video), enabling the development of creative applications ranging from online games to virtual museums and galleries. These documents also can be linked to other pages so that users can easily access more information.

market for innovators. A problematic issue with regard to Web content is how to manage it all. How does one assess and even cope with the amount of information available? How does one access what one wants? How do people know when information is valid? Will people rely on information retrieved for them, or will they still want to develop internal knowledge to be used for judgment? How does one redefine and control search engines? How do people ensure that search engines are objective without being trapped by gluey sites? How do people filter the results of searches to suit personal needs? New cultural models of editorial sense-making have been defined technically and to a certain degree tested: for example, collaborative filtering, "slash-dot"-style social intelligence applied to a common editorial base by combining a group rating, the social status of content-evaluators, and user interface design. These approaches are good examples of hybrids requiring strong arts and humanities skills in the content developing, shaping, rating, expression, and design, and strong IT to deliver effective new platforms sensitive to the community-content co-evolution.[40]

The engagement of artists and designers in the design discussions for new technologies can help ensure that the technologies are as powerful and thoughtful as possible—which is in the interest of everyone (see Box 3.6 for one possibility). However, engagement is a two-way street: Technologists need to be receptive to the input from artists and designers, and artists and designers must take the initiative to become engaged with technologists (e.g., by participating in mainstream technical conferences—and not only those related to animation or graphics).

## ECONOMIC REALITIES

A key factor that shapes the development of tools for creative work is the economics of software. Software is expensive to develop and inexpensive to distribute. Therefore, it is most profitable when it appeals to many people, especially to non-experts (including the mass market), and competition places a premium on being first to market, another factor driving design compromises. The widespread use of a software application can provide a return on the investment required to create complex software systems. Of course, many hardware and software tools developed initially for niche groups eventually find their way into much broader, usually unanticipated, markets. Indeed, some of the best-known information technologies—from the spreadsheet to the Web—were designed initially for narrow purposes. But niche groups, including artists, have little luck in demanding special capabilities—even when they have good connections to vendors. For

> Software is expensive to develop and inexpensive to distribute. It is most profitable when it appeals to many people, especially to non-experts.

---

[40]There is a well-developed, relevant literature in information retrieval, information filtering, information seeking, information organization, and related areas.

## BOX 3.6
### Databases and Artists

Databases touch nearly every aspect of daily life. They present opportunities that artists are only beginning to explore.[1] Countless activities, such as using a credit card, visiting a physician, withdrawing cash from a bank, or accessing Web sites, involve a database that mediates the transaction. The operation of these databases—how (your) information flows—is largely invisible to the general public. One function that art can perform extraordinarily well is to make things visible. A cookie, for instance, is merely data flowing between computers. Imagine how an artist might be able to depict this flow to communicate to the general population what exactly is being transferred—and possibly even some of the implications of the transfer.

FIGURE 3.6.1 An artist's perspective on database privacy. Illustration created by Jennifer M. Bishop, Computer Science and Telecommunications Board staff.

[1]See Victoria Vesna, 2002, "Database Aesthetics: Of Containers, Chronofiles, Time Capsules, Xanadu, Alexandria and the World Brain," *AI and Society*, October 29, available online at <http://time.arts.ucla.edu/AI_Society/vesna_essay.html>.

example, a leading photographer who has achieved acclaim in digital and conventional photography has spent years working with corporations such as Adobe, Apple, Canon, Epson, and Kodak but reports that there are limits to what the software engineers understand about his artistic needs. He attends seminars with vendor engineers and asks for features he needs for his art making. He teases that the problem might be too hard for the engineers, hoping to harness their pride, recognizing that his goal is to get a new feature that is needed by only a very few users.[41]

Non-profit institutions and cooperative groups develop some of the best tools for creative use—both in computer science and in artistic and design contexts. This is true to a large extent because this software is simply not viable commercially and does not appear in that sector. Many good tools are non-commercial and open-source (see Chapter 7), developed and maintained by a number of programmers who coordinate their work through the Internet. This approach allows users to contribute capability directly, makes the software freely available, and allows users access to the inner workings of the software so that creative modifications and customization are possible. See Box 3.7. But it is not a panacea: Open-source tools tend to depend on volunteer efforts and often reflect the design effort of one or just a few primary implementers. As with closed-development/proprietary commercial products, users may not always get the features or design changes they need. Software robustness, documentation, and support may not reach commercial levels.[42] Support and documentation tend both to benefit from and to attract a larger user community. Small, specialized projects often have the feel of "work in progress."

Subsidies are a way to encourage the development of creative tools. This approach is exemplified by the Studio for Electro-Instrumental Music (STEIM), a center in Amsterdam that helps many musicians create new instruments using sensors.[43] By subsidizing the engineering costs of sensor systems, STEIM has enabled the exploration of many novel electronic musical instrument designs. The engineering resources established by STEIM have attracted artists from around the world who otherwise would have no access to this technology. This approach seems to resemble that seen in computational science in the 1980s, when centers providing access to specialized computing resources were created. More generally, government (or philanthropic) support for research, including the development of prototypes and testbeds (see Chapter 8), is a way of subsidizing the development of creative tools that the market, all other things being equal, is not likely to develop in a timely manner. Whether induced through market

> Non-profit institutions and cooperative groups develop some of the best tools for creative use.

---

[41]Personal communication between the photographer and a reviewer, summer 2002.

[42]There are important counter-examples such as the Linux operating system, which is often used in critical server applications.

[43]For further information, see <http://www.steim.org/steim>.

---

**BOX 3.7**
**Open-Source Tools**

*Tools That Support ITCP Work*

Python is a programming language created in the early 1990s by Guido van Rossum, who is still its principal author. Python has gathered a large following, finding important applications in graphics including Alice (see below), gaming, Web services, teaching, collaboration support, and research. Python is copyrighted but is free for use and distribution, and versions are owned by the Centrum voor Wiskunde en Informatica (National Research Institute for Mathematics and Computer Science in the Netherlands), the Corporation for National Research Initiatives, and BeOpen.com. In 2001, Zope Corporation and others sponsored the formation of the Python Software Foundation, a non-profit corporation set up to own the intellectual property of the most recent versions of Python. Conferences and workshops devoted to Python have been held since 1994, and in 2002, Python conferences were held in both the United States and Europe. Python users actively develop extensions and support for Python, including graphical programming environments and debugging tools, a re-implementation in Java (another popular programming language), and many new modules that support such things as Internet access, the XML language, linear algebra, encryption, and multimedia.[1]

PortAudio is an example of a much smaller open-software effort. PortAudio is a cross-platform library that supports audio input and output. It began with discussions about the need for such a library on an Internet mailing list, music-dsp.[2] Hosted by the California Institute of the Arts, PortAudio was developed primarily by Ross Bencina and Phil Burk, who jointly control its design and distribution. More than 30 contributors have fixed bugs, offered expertise on the use of specific systems and hardware, and adapted PortAudio to new systems and devices. A dozen or so commercial and non-commercial applications are implemented using PortAudio.[3]

In both cases, it is interesting that many of the developers and contributors have never met face-to-face. Designs and work plans are discussed via e-mail and the community is informed of recent developments,

---

[1]See <http://www.python.org> for more information.
[2]See <http://shoko.calarts.edu/~glmrboy/musicdsp/whatis.html>.
[3]See <http://www.portaudio.com> for more information.

---

forces or supported by other resources, the research in directions discussed here for hardware and software can be a powerful investment in the future.

## STANDARDS

A review of hardware and software would be incomplete without touching on the perennial issue of standards, which can both limit and support creativity. They obviously establish some constraints, but they also can allow various programs to interoperate in new and creative ways, and they are typically associated with stabilizing a market, lowering costs, and facilitating training.[44] Although perhaps

---

[44]This situation is exemplified by consumer electronics, but it also applies to various forms of IT (and non-IT products).

problems, and solutions through electronic bulletin boards. Most of the contributors work without monetary compensation. In both examples, it is important to have a relatively small group to coordinate and integrate various contributions in order to maintain a coherent product.

*Open-Source Tools for Learning*

Python, based on early Lisp-like languages, is excellent for art and design education because of its simple and powerful syntax. An example of looking ahead to new models for languages is the Design by Numbers project at the Massachusetts Institute of Technology,[4] which includes a simple language for graphical expression (called DBN), distributed freely over the Web, as a means to teach programming to visual artists and designers. This system is currently in use in more than 25 schools worldwide, with a sequel called the Proce55ing project that introduces two-dimensional and three-dimensional graphics concepts to visual designers.[5]

Individual tools available online include Alice,[6] a three-dimensional (3D) interactive graphics programming environment for Windows built as a public-service project by a research group at Carnegie Mellon University. The intent was to make it easy for novices to develop interesting 3D environments and to explore the new medium of interactive 3D graphics. Alice is primarily a scripting and prototyping environment for 3D object behavior, not a 3D modeler, but it does read many common 3D file formats. By writing simple scripts, Alice users can control object appearance and behavior, and while the scripts are executing, objects respond to user input via mouse and keyboard. The current version of the Alice authoring tool is free to everyone and runs on computers that are commonly available for reasonable prices. Worlds created in Alice can be viewed and interacted with inside a standard Web browser once the Alice plug-in has been installed. The Alice core distribution includes a large library of textured models. The Alice plug-in does not allow authoring, but it does allow viewing of any Alice world.

---

[4]Led by John Maeda at the MIT Media Lab. See <http://dbn.media.mit.edu>.

[5]Proce55ing, an electronic sketchbook for developing ideas and a context for learning fundamentals of computer programming within the context of the electronic arts, was initiated by Ben Fry and Casey Reas at the MIT Media Lab and by the Interaction Design Institute Ivrea. As of November 2002, the software is in a pre-release stage, but bug fixes are being made in preparation for a more complete 1.0 release. Proce55ing will be free to download and use. See <http://proce55ing.net/>.

[6]See <http://www.alice.org>.

more obvious in IT, standards are common in arts and design arenas, from the dimensions of a violin to the Pantone color system. Standards can emerge from a standards-setting body or de facto from a leading supplier (as a consequence of competition within the marketplace). A good example for ITCP is digital images and associated software. Digital images are basically just arrays of pixels, yet they can be produced, stored, and manipulated in many ways. With standard representations for digital images, many programs and devices can interoperate, including cameras, digital editors, Web browsers, optical character reading (or other scanning) software, and graphical interface tools. But what should be standard? An image can exist in several forms—a set of high-level instructions for rendering, a set of polygons, or an array of pixels. A pixel representation may be a universal standard, but it forces the loss of high-level information. A high-level standard may be more complex to define, and to agree to, but it can preserve the ability to perform high-level operations. Another example is the musical instrument digital interface (MIDI) standard. MIDI was created to connect music keyboards to synthesizers, but it was soon realized that MIDI could allow computers to control synthesizers. The availability of low-cost synthesizers that are more or less interchangeable has been critical to any number of creative computer music systems.

The rise of ITCP raises questions about the standards-setting process and the opportunities for artists and designers, as users, to have input. It is not uncommon for user groups to be on the sidelines of vendor efforts to set standards; artists and designers are not being singled out. And as recent attempts by non-profit organizations to explore how to inject "public interest" voices into IT standards-setting demonstrate, it is not easy for people with less technical depth, fewer financial resources, and different conversational norms to participate effectively in various IT standards-setting cultures.[45] That said, a constructive step would be for specific communities of practice within the arts-design/ITCP world to try to determine for themselves a consensus view of the kinds of features or qualities they would like to see. A consensus statement would carry more weight with standards setters than would input from lone individuals or representatives of small or niche groups. As suggested above, research prototypes could aid in fostering consensus by demonstrating what a standard may lead to in practice.

*Research prototypes could aid in fostering consensus by demonstrating what a standard may lead to in practice.*

## SELECTED AREAS FOR THE DEVELOPMENT OF HARDWARE AND SOFTWARE THAT WOULD PROMOTE CREATIVE WORK

Precisely what the future holds is uncertain, but based on current expectations and trends,[46] the capabilities available for work in ITCP will become far more powerful and diverse in the coming years. The timing of the present report is therefore propitious. Among notable research trends, nanotechnology is becoming a reality.[47] Computers,

---

[45]For example, the advocacy group, the Center for Democracy and Technology, has tried an experiment in the last couple of years to see how it might participate in Internet Engineering Task Force (IETF) standards-setting activities (John Morris, a lawyer, was the group's participant). The Ford Foundation also fostered discussion of standards setting in its Digital Media Forum. Of course, since the early days of computing there have been various user groups that tried to influence the product designs of major vendors.

[46]Raw computational power has been rising exponentially for years. A popular yardstick is Moore's law: For the last 40 years, computing capability per dollar has doubled every 18 to 24 months, equivalent to a 100-fold improvement every 10 to 13 years, reflected in both rapidly increasing performance and declining price. Internet traffic has been growing by a factor of about two annually, outstripping the growth in computing speed.

[47]Nanotechnologies are generally defined as having structures in the size range of 1 to 100 nanometers (billionths of a meter). Technologies of this size often exhibit novel properties. See, for example, information on the National Nanotechnology Initiative at <http://www.nano.gov/>.

sensors, and other devices are becoming ever smaller. New sensors for sound and light combined with faster and cheaper digital signal processors will make large-scale system sensing increasingly practical.[48] The synthesis of materials to meet defined requirements means that the manufacture of devices will be simplified. In the near future, a number of specific changes are envisioned. Computing speed will become so great (10 gigahertz in a handheld device) that operations as complex as real-time voice recognition without a learning curve could be commonplace. There is also likely to be enhanced machine intelligence and capability for seeing, hearing, and speaking.[49] And many devices will likely be considered disposable.

Thus, although some excellent tools for ITCP work are available, there are many opportunities for improvement. The committee identified a handful of areas that could exert considerable leverage in promoting ITCP, because such tools often enable the widespread adoption of new technologies.

## Distributed Control

One area in need of support is distributed control. Software that coordinates multiple computers is among the most difficult of programs to write. This is true even for commercial business systems, but the problem is more acute in creative contexts because programming resources tend to be much more limited there. Tools can provide simplified protocols for communication between networked computers, including wireless devices.

The essentially sequential nature of software systems has made it difficult to envision a plausible distributed control paradigm. Much research in the related area of parallel computation was curtailed in the 1990s due to the evolution of the conventional microprocessor from the megahertz to gigahertz speed range in only a decade—there was no need to make computer programs run faster when the computer itself had become faster in a matter of months. Distributed control has taken center stage once again, not as a potential means to increase the speed of computation but as a way to handle the difficult task of coordinating computation across multiple computing units. There is a simple impediment to advancing the state of the art in this area; namely, programming methodologies for asynchronous computing are not in place, even in high-end systems. Software system designs inherit a legacy similar to that of the automobile industry in that only incremental, evolutionary changes are adopted instead of the kind of revolutionary advancements needed to write more sophis-

---

[48]From Computer Science and Telecommunications Board, National Research Council, 2001, *Embedded, Everywhere: A Research Agenda for Networked Systems of Embedded Computers*, National Academy Press, Washington, D.C.

[49]See, for example, John Markoff, 2002, "Technology Gives Sight to Machines, Inexpensively," *New York Times*, June 17, available online at <http://www.nytimes.com/2002/06/17/technology/17VISI.html>.

ticated programs that take advantage of multiprocessor and multisystem interactions.

## Sensors and Actuators

Another area deserving attention is sensors and actuators. Artists and designers and scientists alike often want to collect data in the real world and to control mechanical devices using computers. Advanced sensing technologies and actuators exist, but without tools to simplify their use, these technologies remain out of reach of most individuals who are not specialists. Sensors and actuators that can be used by non-experts can promote creative endeavors (see, for example, Box 3.8). Note that the area of motion control and robotics is experiencing a renaissance in computer science: Advances in component technologies and artificial intelligence moved the field along considerably in the 1990s, and research in the area is expected to thrive.

A subtle reason for recent advancements in this area can be associated with the development of the Web. Once-difficult-to-obtain catalogs or obscure parts listings that were available only to hacker communities are now easily found on the Web with a quick search. Interfacing with these parts was once also difficult, requiring access to the relevant technical community. Now most of this knowledge is published and easily accessed on the Web. Interfacing information, sample projects, and even advanced research can easily be found by the curious builder.

In addition to the challenge of using sensor and actuator technologies, there is the problem of interfacing them to computers, especially if they are non-standard. This situation might be alleviated by the design and distribution of software and hardware that take advantage of serial interfaces such as the universal serial bus (USB) and the IEEE-1394 multimedia standard.[50] There have been several attempts at developing general-purpose sensor systems and input/output boxes in the arts, such as the iCube system and notably, Michael Rodemer's EZIO board, designed at the Art and Technology Department of the Art Institute of Chicago in the mid-1990s and now commercially available.

The success of the Parallax Inc. Basic Stamp microcontroller as a low-cost platform for educational experimentation has changed the public's access to embedded computing. Ample documentation, complete technical information, and well-documented sample projects have

---

[50]One interesting effort is the development of "phidgets," or physical widgets, at the University of Calgary. Phidgets are building blocks that assist in the development of physical user interfaces. Phidgets consist of a hardware device, software architecture for communication and connection management, a well-defined software application programming interface (API) for device programming, a simulation capability, and an optional component for interacting with the device, as described in "An Overview of Phidgets," available online at <http://www.cpsc.ucalgary.ca/grouplab/phidgets/>.

made this the platform of choice for many budding "physical comput-ing" courses nationwide. Once one is versed in the Basic Stamp microcontroller, it is easy to move to more economical systems by going one level deeper into the inner workings of the Basic Stamp to its central controller, the "PIC" chip. A wide variety of PIC chips, and competitor single-chip computers such as AVR Inc.'s line, have made it possible for anyone to easily build a sub-$10 computer for embed-ded control. Any art education program should find it attractive that platforms priced at less than $10 can be budgeted for as easily as any other art supply like paint, canvas, or a brush.

With respect to interfacing the various devices that are emerging today, one challenge is the speed and reliability of data transfer. Re-cently buses such as the CAN bus have emerged as high-speed stan-dards for the robotics community, and the consumer-class USB and IEEE-1394 technologies are continually giving rise to faster versions like USB-2 and Firewire-2.[51] Yet at some point the rapid progress in these standards can lead to confusion in the general population. What economists call the cost of switching—having to relearn a technology that one had finally felt comfortable with but then learned had sud-denly become passé—can deter or slow adoption of technology, as well as influence standards-setting tactics.

## Video and Audio

Extensive editing and processing tools exist for time-based media (including video and audio), but these tools assume a particular style of working—namely, that the product will be a linear video or audio recording. In addition, most tools do not support real-time processing. These limitations hamper artists who want to work with video and audio media in interactive performances, take other approaches to interactivity, stream material over the Internet, or use scripts to pro-cess or generate video.

Standards continually emerge and compete in this arena such as Apple QuickTime, Macromedia Flash, and Real Video. A common difficulty for all of these standards is to maintain compatibility across the multiple platforms they support. Often the same version of any system, such as Flash, will display a file properly on a PC but not on a Macintosh. These inconsistencies are rarely addressed, and content creators must simply work around these known problems. The stan-dards are constantly upgraded, solving some problems yet introduc-ing others. For an emerging field, like digital media or ITCP generally, standards that lack reliability are a serious problem. It is as if the art of film were emerging in an era when the film you might have shown just once might suddenly dissolve before your very eyes.

---

[51]Buses enable data transfer among the different components of a computer.

## BOX 3.8
### Ghostcatching

The fruit of a collaboration between multimedia artists Paul Kaiser and Shelley Eshkar and dancer/ choreographer Bill T. Jones, Ghostcatching[1] is a virtual dance performance that makes use of motion-capture technology. Envisioning a blend of performance, filmmaking, drawing, and computer composition, the artists used light-sensitive sensors attached to 22 key points of Jones's body and eight cameras to capture his movements. Roughly 40 sequences of Jones's movement were recorded digitally. The resulting "capture data"—which represented a record of only the movements of the sensors and not of Jones's body per se— were then used as the raw information for Kaiser and Eshkar's creative application. See Figure 3.8.1. Jones's "movements were then manipulated electronically and re-choreographed on a computer screen to make an original virtual performance."[2]

The results include an 8-minute digital projection, 13 still images taken from the dance, photographs that describe the artists' process, and a soundtrack consisting of Jones's own sounds: "chanting, humming, singing, talking, grunting, and more."[3] Commissioned by the Cooper Union for the Advancement of Science and Art,[4] the installation opened at the Cooper Union's Arthur A. Houghton Jr. Gallery in New York City and ran from January 6 to February 13, 1999.

1. Improvising

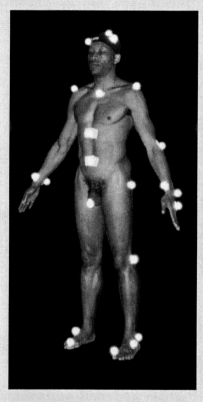

2. Wearing motion-capture markers

---

[1]For more information, see <http://www.cooper.edu/art/ghostcatching/ghost/exhibit.html>.
[2]Zoe Ingalls, 1999, "Using New Technology to Create 'Virtual Dance'," Chronicle of Higher Education 45(21): A29-A30.
[3]Ingalls, 1999, "Using New Technology to Create 'Virtual Dance'."

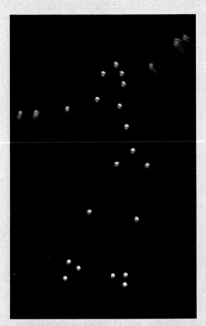

3. Markers optically recorded and converted to digital three-dimensional files

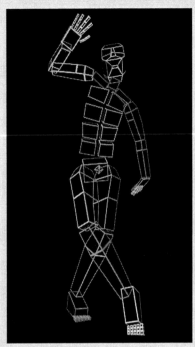

4. Motion files applied to kinematic model of the body

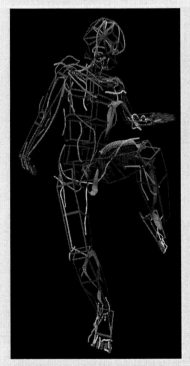

5. "Hand-drawn" lines modeled as mathematical curves

6. Sampled charcoaled strokes applied and rendered as a final drawn body

FIGURE 3.8.1 Ghostcatching. Images courtesy of Bill T. Jones, Paul Kaiser, and Shelley Eshkar; available for download at <ftp://ftp.eshkar.com/ghostcatching/>.

## Generative Processes

From both a computational and an artistic and design standpoint, an attractive feature of computers is that they enable generative processes. Rather then creating each note or brush stroke by hand, an artist or designer can design a process that generates material automatically. Of particular interest are fractals, chaotic systems, and systems exhibiting non-linear dynamics. These have strong connections to natural processes, can be explored only by using computers, and often create interesting and unpredicted output. A programmer could set up a generative process relatively easily, but tools are needed that can encapsulate these processes to make them more accessible to artists and designers. Generative processes are interesting to contemplate for a few reasons. First, the associated computer models depend on assumptions—and to ensure that the models do not present the kinds of constraints criticized above, it will be important for artists and designers to have some understanding of the models' inner workings, which in turn depends on comfort with quantitative analyses. Second, some of the earliest explorations of ITCP were generative processes. A notable instance is Harold Cohen's work with the artificial intelligence drawing-generating program, Aaron.[52] More contemporary simulation models, some of which explore "artificial life," have been used by biologists and other scientists to understand evolution, yielding a new understanding of its pacing and realization. The Santa Fe Institute led the development and use of these models, providing training in modeling methods, and it also has reached out to the arts community. The rise of computer games that feature artificial worlds suggests but one of the more obvious potential ITCP connections. Artificial-world development is also being explored outside the game context.[53]

When computing was young, the most natural form of creative activity on the computer was programming. This led to some landmark work by researchers at Bell Labs, such as Ken Knowlton, who were versed in programming but in addition had a penchant for visual art. The idea that forms can emerge from programming echoes complex processes in the physical world where the result is something you cannot expect. More recently with the rise and proliferation of digital tools, programmatic approaches to form have become less popular. For ITCP to advance, a fundamental shift is needed from computation or mathematics based on numbers to that based on symbols.

---

[52]See <http://www.umcs.maine.edu/~larry/latour/aaron.html>.

[53]See the World Generator/The Engine of Desire by Bill Seaman with Gideon May. The work enables the construction of virtual worlds in real time in an interactive environment. Although the work is artistic, it has been featured in a number of IT studies and has potential for many design and interactive tool applications. For further information, see <http://faculty.risd.edu/faculty/bseamanweb/web/texts.html> and <http://www.nada.kth.se/erena/pdf/D1_2.pdf>.

## Reliable, Low-latency Communication over the Internet

Artists and designers have common cause with a wide range of users in their interest in reliable, low-latency communication over the Internet.[54] Higher bandwidth and/or quality-of-service guarantees could enable new levels of interaction, distributed concerts, two-way video, and other creative activities that necessitate specific levels of network performance. As some observed to the committee, were artists able to stream their own bandwidth-demanding work out, it would yield greater diversity of voices and stimulation to creativity. This is the crux of the broadband dilemma: Broadband users are twice as likely as dial-up users to have their own Web sites, with nearly 60 percent becoming creators and managers of content in this manner, according to a report by the Pew Internet & American Life Project.[55] But pervasive broadband deployment in the United States will require billions of dollars in investment;[56] subscriber levels, while growing, fall far short of the potential, given existing deployment, and therefore widespread use is not expected soon. That said, in the academic world, the Internet2 project and complementary federal research infrastructure programs provide high-bandwidth service on and between many campuses because of its potential to support cutting-edge application research. Music performance over the Canadian equivalent, the CANARIE network, illustrates the general proposition that these facilities could be valuable ITCP vehicles. Recent advances in low-latency ultra-videoconferencing have been carried out by musician-researchers at McGill and Stanford Universities.[57]

As in other communities, opinions in the arts world about the value of broadband differ. To some, a benefit of ever-increasing processor speeds has been the development of media-rich simulations and other works, stimulating demand for broadband services. Yet others find such digital spectacles less compelling than new social models for peer-to-peer cultural production. From this standpoint, advanced content should be focused not on creating volumes of traffic

---

[54]The Internet is based on best-effort policies that make no guarantees for the timely delivery of information. Reliability is achieved by re-transmitting data when a loss is detected, but this recovery introduces delays. See Computer Science and Telecommunications Board, National Research Council, 2001, *The Internet's Coming of Age*, National Academy Press, Washington, D.C.

[55]Mike Snider, 2002, "Faster Connection Allows Users to Do More," *USA Today*, June 23, available online at <http://www.usatoday.com/life/cyber/tech/2002/06/24/broadband.htm>. See also *The Broadband Difference: How Online Americans' Behavior Changes with High-speed Internet Connections at Home*, available online <http://www.pewinternet.org/reports/toc.asp?Report=63>.

[56]See Computer Science and Telecommunications Board, 2001, *The Internet's Coming of Age*, p. 45.

[57]Jeremy Cooperstock, an engineer at McGill University, and Chris Chafe, a composer at Stanford University; see <http://www.cim.mcgill.ca/~jer/research/rtnm/>. Internet2 and CANARIE are discussed further in Chapters 5 and 8.

A quantitative

increase in bit

traffic makes

possible

qualitative

change in digital

expression.

that can fill up fat data pipes, but on infrastructural advances that would allow a wide public access to and control over digital objects, although with the recognition that there are certainly legitimate and productive needs for greater capacity.[58] This perspective shifts the focus to thinking of how to increase the range of interpersonal connection and personal expression distributed across society.

One issue relating to broadband is that the quantitative increase in bit traffic makes possible qualitative change in digital expression. However, this potential is confounded by a tendency for capabilities to fluctuate with IT innovations: After the initial emergence of digital media serving smaller amounts of kilobytes, higher amounts of data could be conveyed through the CD-ROM media, only to go back to smaller amounts of kilobytes over the (narrowband or dial-up) Web, and then up to large capacity again via DVD-ROM and now to broadband. Looking across this series of innovations, they have not been associated with an intrinsic change in content for the better—perhaps because of uncertainty associated with the fluctuations, or perhaps because experience with any one technology has been too abbreviated.

## Tool Design and Human-Computer Interaction

There is a great deal of work to be done in understanding the ramifications of the tradeoffs that occur every day in the design of tools. The discussion above includes examples of how the tool designers' choices in areas such as abstraction, extensibility, and metaphor can affect the practices within a field and the results obtained through the use of the tool. Insufficient work has gone into understanding how to leverage this relationship to better support creative practices. This goes far beyond the usability studies once typically associated with human-computer interaction research. Questions to be addressed include these: When does embodying domain knowledge in a tool help? What is the impact of supporting multiple metaphors in one tool? When are constraints good inspirations, and when do they cause excessive perspiration? When can implicit assumptions help, and when can they be hindrances? How much and what kind of openness yields opportunity, and what yields frustration? In other words, what is the relationship among information, its representation, presentation, and manipulation, and the range of work encouraged? Researchers and developers have produced a small number of data points—more work in understanding this area can have deep ramifications for encouraging ITCP.

---

[58] A Canadian task force concluded a series of consultations with the arts and cultural community with a report entitled *Filling the Pipe: Stimulating Canada's Broadband Content Industry Through R&D*, available online at <http://www.canarie.ca/press/publications/broadband_report.pdf>.

## *Programming Languages*

Finally, some areas of great interest to artists are not well served by programming languages, which tend to be driven by science, engineering, the Web, and more commercial application areas. Some of the topics discussed above, especially distributed control, video and audio processing, and generative processes, could be supported better by languages. Languages are particularly important for specifying interactive behavior and generative processes, an area not well supported by commercial software. Because a great deal of creative activity involves combining existing concepts in new ways, and because programming languages provide the glue for assembling software tools and libraries into applications, languages are critical to innovation. Progress has been achieved in making programming languages simpler for novice programmers, but this work needs to be adapted to support more creative work.

# 4 The Influence of Art and Design on Computer Science Research and Development

nformation technology (IT) as a medium for the work of artists and designers is discussed in Chapter 3, which points out that there are many ways for computer science (CS) to support new tools and applications for the arts and design disciplines, in service to cutting-edge and more mainstream practitioners alike. These tools and applications offer the potential for beneficial developments in information technology and creative practices (ITCP). But there are further, more profound implications of the intersection between IT and the arts and design, and these are the focus of this chapter, which views art and design practices as forms of CS research and development. This perspective on CS is more subtle, more challenging, and more fundamental than the tools orientation of Chapter 3. It involves a non-traditional and perhaps unfamiliar kind of art and design practice. It also involves rethinking CS in ways that many computer scientists would find nontraditional.

## BEYOND TOOLS

### THE INFORMATION ARTS

Writing in 1993 during the take-off of the wired boom of the 1990s, veteran commentator Stewart Brand pondered whether "technology has swallowed art, and so is art gone now?"[1] In fact, if art is understood as the making of unique individual objects—such as paintings, sculptures, and drawings—or the result of traditional approaches to

---

[1]Stewart Brand, 1993, "Creating Creating," *Wired*, 1.01 March/April.

the performing arts, then, for some new-media artists, the answer may be yes. They take information technologies for granted, but their art is not fixated on the computer as a medium, as if it were paint or a violin, or even the sound artist's turntables or the scenic artist's optical instruments. As Stephen Wilson's recent encyclopedic compendium of contemporary intersections between art, science, and technology shows, the *information arts* range across the life and space sciences, nanotechnology, robotics, and other new materials, as well as IT itself.[2] This style of practice does not use technology to create new artworks so much as it uses artistic practice to manage and interpret information at the cusp of technological and scientific research.

This new kind of art and design practice looks increasingly like technical research, but it is done from an artistic or design rather than a scientific perspective—it asks different kinds of questions and uses different kinds of methods to search for answers. Generally speaking, technical research focuses almost exclusively on new technical possibilities: What new things can be done? How can they be done faster or more efficiently? By contrast, artistic and design work tends to focus on the social and cultural meaning of the technology that is under development. This aspect differentiates the approach from that of conventional CS, which does not tend to address explicitly such implications of decisions about system design and implementation, and which may look askance at approaches that have a social science flavor. While a traditional work of art can be thought of as a representation of an artistic concept, the information arts often ask what technologies themselves (perhaps unintentionally) express and how they ought to be reconceived.[3]

Artists' questioning can be a powerful, constructive force. In particular, since the mid-19th century artists have often personified the "user to come" for new cultural technologies. Many media technological advances have arisen in the arts and design fields or have been modeled there, a decade or a generation ahead of the industrial-academic curve; see Box 4.1. For Alvy Ray Smith, the prominent computer graphics expert, artists are most valuable as "explorers at the edge of our culture," and he looks to them to "tell the rest of us what [computation] really is."[4] Thus, the information artist functions as an archetypal knowledge worker: someone able to "penetrate conventional organizations to which their continuing attachment to an 'external' knowledge community represents a valuable asset."[5] ITCP

---

[2]See Stephen Wilson, 2001, *Information Arts*, MIT Press, Cambridge, Mass. Also see links to online resources at <http://online.sfsu.edu/~infoarts/links/wilson.artlinks2.html>.

[3]The CAT's MeAoW lecture series at New York University, for example, was framed around such questions; see Box 6.1 for details.

[4]Alvy Ray Smith, 1998, "The Stuff of Dreams (25 Years = 100,000x)," *Computer Graphics World* 21(7): 27-29.

[5]See Paul A. David and Dominique Foray. 2002. "An Introduction to the Economy of the Knowledge Society," *International Social Science Journal* (UNESCO 171) 54:25-37.

**BOX 4.1**

**Far from Unprecedented:**

**Influences of Art and Design on Information Technology**

The role of artists in defining technology is well established in history, though this history is often not well-known, as least by modern pundits and chroniclers of information technology (IT). For example, Louis Daguerre, often cited as one of the pioneers of photography, was neither a chemist nor an optics professor; he was a painter of sets for the opera (and the inventor of daguerreotype).[1]

It is instructive to consider the history of electronic and digital musical instruments. From the point of view of composers, the critical discussions were about timbre, scales, and compositional techniques and influences, thus shaping how electronic and digital music instruments were imagined, used, and developed. The instruments, performances, and means of distribution were all co-developed between artists and engineers.[2] The artists' work was a precondition to commercial exploitation (in the direct and indirect senses).

Such a process is contrary to the traditional historical accounts of engineers (or the military) driving the development of technology and the artists following by finding "creative applications." Instead, it is often the artists who make the initial investment. The investment of artists is critical to this innovation, as it is artists who are motivated to explore alternatives beyond what has already been framed as acceptable, often long before major commercial applications can be imagined (e.g., the Hammond organ is the direct application of patents filed by a musician before the miniaturization of electronics).

Why are such observations on the way to innovation often omitted from histories of technology evolution? First, art critics often do not "do technology"; art critics and curators, because of their education and expertise, generally do not have much acquaintance with technological concepts, technologies, or theoretical tools for techno-social interaction. Second, because commercial success has become the dominant measure by which to evaluate and justify technological and scientific research, less attention is accorded systematically to non-commercial applications.

---

[1]See "Daguerre, Louis Jacques Mande" at <http://www.rleggat.com/photohistory/history/daguerr.htm>.

[2]See David Dunn, 1992, "A History of Electronic Music Pioneers," an essay written for the catalog accompanying the exhibition "Eigenwelt der Apparatewelt: Pioneers of Electronic Art," Ars Electronica, Linz, Austria, curated by Woody and Steina Vasulka. Also see Trevor Pinch, 2002, *Analog Days: The Invention and Impact of the Moog Synthesizer,* Harvard University Press, Cambridge, Mass.

---

**The reach of the information artist extends beyond product design to process design.**

makes apparent the value of the artist as mediator—someone who is increasingly intercommunicating—addressing IT-related process and context and expanding beyond the traditional artist's focus on content.

For an example from the world of design practice, recall the work of Karim Rashid (presented in Chapter 2). The software that controls the variations in each of the napkin rings produced is integral to the creative process. Without such software control, the rings would be identical as the outputs from a mass production process. In both instances (with and without the intervening software), the initial design involves creative practice. However, the introduction of Rashid's software into the production process offers an additional opportunity for creativity. In this sense, the reach of the information artist extends beyond product design to process design.

## MODELING DISCIPLINES: FROM MULTIDISCIPLINARY TO TRANSDISCIPLINARY

The relationship between IT and the arts and design as discussed in Chapter 3 can be described using a *multidisciplinary* model. Each discipline (e.g., architecture) is represented as a circle; the area where the circles overlap is the area of intersection (e.g., IT-enhanced architecture). One implication of this representation is that the non-overlapping areas do not change. Each discipline provides some piece of its practice or theory that is compatible, useful, or mutually beneficial to other disciplines without any change in the way the discipline itself fundamentally works (see Figure 4.1a).[6] This kind of representation conveys how aspects of already-existing genres, methods, or theories are *applied* in different contexts—in the case of this report, how IT can be applied in the arts and design areas.

But in another possible model (Figure 4.1b), the circles do not intersect but instead share a common frame. Each discipline maintains its own knowledge and methodologies but is fully, not partially, open to the other disciplines.[7] In this second model, the disciplines not only apply their methods in a new context but also are receptive to fundamental changes in knowledge and methodology based on their interaction. What is crucial to enable this kind of interaction is the "frame" surrounding the disciplines—a mutual awareness and understanding, especially a historical understanding, of one another and their relationship. The shared frame may be only a transient phenomenon—the disciplines may come into contact, engage in some fruitful exchange, and then continue to develop separately and move apart, as contrasted with the multidisciplinary approach sketched in Figure 4.1a. In *transdisciplinary* research,[8] the point is not just application of given methodologies but also *implication*—a result of imagining entirely new possibilities for what disciplines can do.

*In transdisciplinary research, the point is not just application of given methodologies but also implication—a result of imagining entirely new possibilities for what disciplines can do.*

---

[6]It is worth noting that areas such as "the arts" cannot be reasonably contained within a circle implying that what is "in" can be distinguished easily from what is "out" (or even that an "in" versus "out" categorization is useful). The use of a circle to represent disciplines is, of course, a heuristic for purposes of exposition.

[7]Disciplines can be relatively open or closed in various ways. One way to assess whether a discipline is open is through the use of tools from bibliometrics. One can, for example, employ citation analysis to determine the frequency with which a discipline cites articles in (certain) other disciplines. Those disciplines with a higher percentage of outside-discipline citations could be characterized as more open than disciplines with lower percentages. Standards (and representational structures more generally) can also be examined in their role of affecting openness. Standards (e.g., HTML) can serve as wonderful contributions to work at the intersection of disciplines, but at the same time can be the source of resistance to change in the home discipline.

[8]Margaret A. Somerville and David J. Rapport, eds., 2000, *Transdisciplinarity: Recreating Integrated Knowledge*, EOLSS, Oxford, U.K.

*(a) Multidisciplinary Model:*
*New Context of Application*
• Mixes knowledge of disciplines A, B, and C for a common purpose, but each discipline keeps its former shape
• Intersected area creates new context for application of already-existing concepts and methods
—IT as a means to replicate, automate, speed up, and/or reduce the cost of prior analog practices
—Art as post hoc "beautification" of separately conceived functionality styling

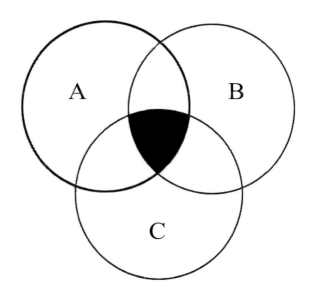

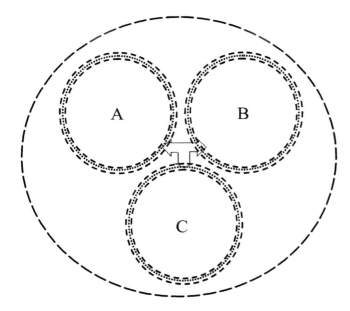

*(b) Transdisciplinary Model:*
*New Context of Application*
Transaction space requires understanding of own and other disciplines
• New common space created specifically from interpenetration of disciplines
• Boundaries are shown as perturbed, and they adjust to accommodate the reflection from other circles
• Intensity of communication between disciplines becomes context for *implication*—expansion beyond context of immediate application to "anticipatory vision" of future possibilities
• Transaction space may be transient, leaving separate circles reconstituted based on a single transaction—or sustained over time, leading to a durable merging

FIGURE 4.1 Models of the relationship between information technology and the arts and design: Multidisciplinary versus transdisciplinary. SOURCE: Adapted from Margaret A. Somerville and David J. Rapport, eds., 2000, *Transdisciplinarity: Recreating Integrated Knowledge*, EOLSS, Oxford, U.K, pp. 248-249.

The project on Internet data sonification (introduced in Chapter 2) involving a Lucent Technologies statistician and a media artist/composer illustrates the differences between the two models. Brought together through an ad hoc, short-term artist-in-residency effort co-sponsored by Lucent Technologies and the Brooklyn Academy of Music, neither the scientist nor the artist had a preexisting hypothesis or conception of how to combine musical gesture and statistical modeling of data. The outcome was a prototype listening station that conveys the dynamic properties of Internet communication flows (e.g., newsgroups and chat communities) through sounds: melody, texture, and rhythm. This common object resulted from reciprocal learning about the implications of each field for the other. Beginning with a less joint or problem definition, a multidisciplinary outcome would have been different from this—and arguably less innovative. For example, a composer or a painter might have interpreted the abstract patterns revealed by the data flows; that kind of approach has been seen in art inspired by data visualization. Or the statistician might have proposed to a designer the production of an elegant display of quantitative tables.

The Listening Post research prototype has spawned several sequels, each realized in the separate worlds of the scientist's and the artist's ongoing work. Mark Hansen, the scientist, defined applications relevant to the operation of Lucent network-monitoring facilities; Ben Rubin, the artist, continued to present data-driven network sonification in musical and gallery contexts. The team has, in fact, endured longer than the initial pilot.[9] Two co-authored papers on the results of the experiment address different specialist readers, accounting for the collaborators' contribution symmetrically.[10] Intellectual property agreements unique to the dynamics of the project were developed after some difficulty; they recognized that the artistic content and the invented intellectual property—together constituting the project outcome—are inextricably merged and are co-owned.[11]

In the intersection of multiple disciplines described above, the roles of artists and designers and computer scientists are clear-cut. Artists and designers have needs that computer scientists can fulfill. Engaging in a fruitful exchange requires conversations to identify those needs and to determine how computer scientists can best fulfill

---

[9]Through the sponsorship of the Rockefeller Foundation, their collaboration continues as of this writing: The "Ben Rubin and Mark Hansen: Listening Post" exhibition is running at the Whitney Museum of American Art from December 17, 2002, through March 9, 2003; see <http://www.whitney.org/information/press/87.html>.

[10]Ben Rubin and Mark H. Hansen, 2000, "The Audiences Would Be the Artists and Their Life Would Be the Arts," *IEEE Multimedia* 7(2): 6-9; Mark H. Hansen and Ben Rubin, 2001, "Babble Online: Applying Statistics and Design to Sonify the Internet," pp. 1-15 in *Proceedings of the 2001 International Conference on Auditory Display*, Espoo, Finland, July 29-August 1.

[11]In 2002, Mark Hansen joined the faculty of the University of California at Los Angeles. Interestingly, Hansen developed two presentations for interviews—one for faculties of science and one for faculties of art and design.

them. In a transdisciplinary situation, however, artists and designers are not clients of computer scientists but instead interact with them as peers. Bringing to the exchange their own disciplinary methodologies and value systems, artists and designers have their own opinions about what research ought to be pursued and how it ought to be done. One result is a fundamental rethinking of how research into information technology might be conceived.

## IMPLICATIONS FOR COMPUTER SCIENCE

For computer science work, the advantages of being open to the perspectives of the arts and design disciplines are potentially large. Computer science already has a productive tradition of drawing on other disciplines, from mathematics to physics to cognitive psychology, to advance its own work by exploring new problems and thinking about new potential solutions to those problems.[12] Similarly, responding to disciplines from the arts and design worlds opens the possibility of discovering new methodologies for and solutions to problems that, until now, have been beyond the reach of the computer science field to solve or perhaps even articulate. Often the effects of IT research have proved profound (and sometimes unintentionally so),[13] and ITCP serves as a way for those with primarily technical interests to communicate with those more interested in the social, cultural, and political aspects of technology.

The perspectives of the information arts are particularly interesting in cases where CS research itself is already moving toward the perspective embodied by art and design practices. One example of such a shift is in the field of human-computer interaction (HCI). In the last 10 years, the field has moved gradually from focusing largely on the hardware and software of human-computer interaction (e.g., development of the mouse and graphical interface) to paying more attention to human psychology (e.g., what mental models of software are constructed by users) and social interaction (e.g., how software can support project collaboration). More recently, HCI has begun to draw more broadly on the social sciences, especially ethnography (the rigorous, qualitative study of human use and contexts of technology), in order to design systems that better fit into the lives of human users. Simultaneously, the connections have deepened between HCI and the design community, which approaches human-computer interaction in more open-ended ways.[14] These shifts in HCI as a field bring it closer

> One result is a fundamental rethinking of how research into information technology might be conceived.

[12]See Computer Science and Telecommunications Board, National Research Council, 1992, *Computing the Future: A Broader Agenda for Computer Science and Engineering,* Juris Hartmanis and Herbert Lin, eds., National Academy Press, Washington, D.C.

[13]For example, it is clear that the developers of Transmission Control Protocol/Internet Protocol were not looking to challenge the music recording industry, or to enable the wide distribution of technical information on making inexpensive bombs.

[14]See, for example, Terry Winograd, John Bennett, Laura De Young, Peter S. Gordon, and Brad Hartfield, eds., 1996, *Bringing Design to Software,* ACM Press, New York, N.Y., and Addison-Wesley, Reading, Mass.

to the information arts and suggest that there is now a potential for synergy between the two.

One such area where the methodology and attention to social, cultural, and political context typical of the information arts may benefit HCI is the use of technology outside of work contexts. Work applications tend to focus on efficient, problem-solving functionality for which we now have well-understood design and evaluation techniques. Applications for everyday life, however, suggest the importance of aspects that are less understood and are hard to quantify, such as quality of experience, meaningfulness, personal values, identity, and appropriateness to social and cultural context—areas for which the perspectives of the information arts may be particularly appropriate. This concern with the human element can already be seen in consumer electronics, which have a history of drawing on market research and human factors analysis and which often depend on design for competitive advantage; the broader uses of IT envisioned with increases in the embedding of computing and communications components implies a broader and often rather different set of personal and other non-work-focused technologies in the future. HCI researchers are realizing that there is a need to do some fundamental rethinking of HCI methods to understand what these assumptions are, to analyze the extent to which they are applicable to the new contexts of everyday life, and, in cases where they are not applicable, to invent new methodologies that are more appropriate.[15] This kind of rethinking is an endeavor for which the information arts can be helpful; concrete examples where interaction may be particularly fruitful are discussed further in the section "Non-utilitarian Evaluation" below.

Similar shifts are occurring in other areas of computer science. In artificial intelligence (AI), for example, there has recently been a focus on lifelike computer characters or believable agents, with a great deal of interest in incorporating approaches from drama and the arts into agent design.[16] The development of algorithms for information retrieval on the Web has underscored the need to combine theoretical

---

[15]See, for example, recent publications of the ACM Press, including the proceedings from the CHI 2002 Workshop on Funology, the HCI 2002 Workshop on Understanding User Experience: Literary Analysis Meets HCI, and the CHI 2003 Workshop on Designing Culturally Situated Technology for the Home; also Bill Gaver and Heather Martin, 2000, "Alternatives: Exploring Information Appliances Through Conceptual Design Proposals," pp. 209-216 in *Proceedings of the CHI 2000 Conference on Human Factors in Computing Systems*; ACM Press, New York; Debby Hindus, Scott D. Mainwaring, Nicole Leduc, Anna Elisabeth Hagström, and Oliver Bayley, 2001, "Casablanca: Designing Social Communication Devices for the Home," pp. 325-332 in *Proceedings of CHI '01*, ACM Press, New York; and Jon O'Brien and Tom Rodden, 1997, "Interactive Systems in Domestic Environment," pp. 247-259 in *Proceedings of the 1997 Conference on Designing Interactive Systems*, ACM Press, New York.

[16]See, for example, Clark Elliott and Jacek Brzezinski, 1998, "Autonomous Agents as Synthetic Characters," *AI Magazine* 19(2): 13-30; and Joseph Bates, 1994, "The Role of Emotion in Believable Agents," *Communications of the ACM* 37(7): 122-125. Also see Box 4.2.

computer science with an understanding of the social structure of the Web[17] and raises potential connections to the cultural politics of Web information,[18] an area in which the information arts are working. For areas like these, in which purely technical solutions do not seem adequate to fully address the problems of interest to computer scientists, interaction and engagement with information arts could be beneficial to computer science.

Because the information arts are inherently transdisciplinary, they hold the possibility of motivating more than just the straightforward use of information arts for computer science's ends or simple collaboration between information artists and computer scientists. Instead, there can be a mingling and repositioning of each interacting discipline. From the perspective of computer science, this implies a move to more qualitative, rather than quantitative, research methods; a greater incorporation of political, social, and ethical considerations into computer science research; and more focus on intuition and aesthetics.[19]  Given the movements that have already taken place on the arts and design side, productive cross-fertilization and a broader base for ITCP will depend on the flexibility and openness of individual researchers, research communities, departments, universities, and professional societies—the institutions and organizations that define academic computer science.[20]  As detailed elsewhere in this report, there is both movement in that direction[21] and resistance to such movement.

## PROMISING AREAS

During the course of its deliberations, the committee identified a number of promising areas for transdisciplinary work. Several have attracted fairly active interest, whereas others are just emerging. They open the possibility of fruitful discussion and collaboration in these

---

[17]See J. Kleinberg and S. Lawrence, 2001, "The Structure of the Web," *Science* 294(5548): 1849-1850.

[18]Richard Rogers, ed., 2000, *Preferred Placement,* Jan van Eyck Akademie, Maastricht, The Netherlands.

[19]For a compelling example of how AI research can be rethought to incorporate critical thinking from the  humanities, see the discussion of critical technical practices in Philip E. Agre, 1997, *Computation and Human Experience,* Cambridge University Press, Cambridge, U.K. For a similar integration of critical thinking and HCI, see Paul Dourish, 2001, *Where the Action Is,* MIT Press, Cambridge, Mass.

[20]See Chapter 6 for discussion.

[21]Sometimes movement may occur in teaching, as opposed to research, the focus of most of the examples in this report. For example, Robert Coover developed "Cave Writing," a course at Brown University that brought together English students, artists from the Rhode Island School of Design, and computer scientists from the Brown CS department. They explored the creative potential of the Cave, a high-end virtual reality environment for text and related digital media elements of sound and image in virtual space.

areas between computer scientists and artists and designers engaged in computer-science-like work. The discussion below focuses on areas that the information arts are particularly well suited to address based on the following factors: The areas involve the social context or politics of computing; they raise difficult ethical issues that need to be addressed in the context of technical research; they have high public or social impact; and/or they suggest fundamental rethinking of computer science. Although the following compilation is neither comprehensive nor predictive of the most promising areas, it does give an idea of the breadth of possibilities for productive engagement between the information arts and computer science.

## MIXED REALITY

Mixed reality is a new, interactive medium in which computing is taken off the desktop or head-mounted display and linked with real-world objects and places to become part of everyday, physical lives. In these approaches, IT development and other creative practices are synergistic. On the one hand, IT provides a new medium for creative expression, opening up a space of possible developments to be explored. Design and media art practice, on the other hand, offer a broader functional and aesthetic perspective. An art and design perspective introduces a cultural awareness that is essential in the development of devices that not only are functional but also contribute to the quality of life in a less direct, but often more profound, way.[22]

In one classic design, Durrell Bishop's marble answering machine, each message is represented by a marble (see Figure 4.2).[23] When a message is taken, the machine produces a marble.[24] The marble can be picked up and put back into the machine in order to play the message. Placed on a matching phone, the marble causes the phone to dial the original caller. Messages are deleted by recycling the marble in the machine. The marble answering machine speaks to humans' physicality.[25]

Approaches to mixed reality include tangible media and augmented reality. In tangible media, physical objects like Bishop's marbles

> For areas in which purely technical solutions do not seem adequate, interaction and engagement with information arts could be beneficial to computer science.

---

[22]Considerations beyond the utilitarian are pervasive and dominant in developed economies: The design and production of myriad products and services, from automobiles and clothing to news services and computer systems, incorporate many features that are not strictly necessary from a functional perspective.

[23]Gillian Crampton-Smith, 1995, "The Hand That Rocks the Cradle," *I.D.*, May/June, pp. 60-65.

[24]The movie *Minority Report* (produced in 2002 by 20th Century Fox and starring Tom Cruise) employs a similar marble to announce the names of the victim(s) and perpetrator(s) of crimes that will take place in the future.

[25]Dag Svanaes and William Verplank, 2000, "In Search of Metaphors for Tangible User Interfaces," *2000 Conference on Designing Augmented Reality Environments*, Association for Computing Machinery, New York.

B. Play messages

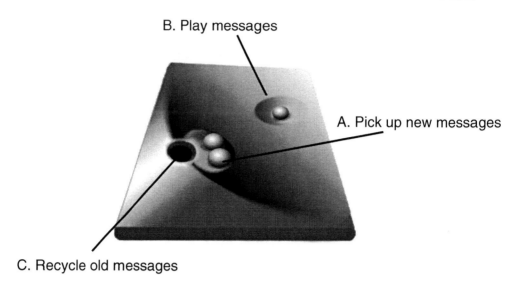

A. Pick up new messages

C. Recycle old messages

FIGURE 4.2 A conception of Durrell Bishop's marble answering machine. When a new message is left, the machine deposits a marble in the upper tray (A) for the recipient to find. The marble can then be placed on the lower indentation (B) to play the message. When the message is no longer needed, the recipient recycles it by dropping the marble into the hole (C). Illustration created by Jennifer M. Bishop, Computer Science and Telecommunications Board staff.

have computational properties. Augmented reality is an alternative to virtual reality in which virtual images and data are projected onto and thereby incorporated with the physical world. For example, one can look through augmented-reality binoculars mounted in the atrium lobby of the Center for Art and Media (ZKM) in Karlsruhe, Germany, to see the heart of the building, overlaid with labels explaining what is done on the different floors: reality plus.[26] Both augmented reality and tangible media have their roots in Mark Weiser's vision of ubiquitous computing.[27]

Technical issues in mixed reality include the maintenance of correspondence between real-world and virtual objects, standards for interobject communication, perception (including vision processing, video tracking of objects, plan recognition, and integration of multiple forms of sensory data), spatial reasoning, and learning and adaptation. But designing and constructing mixed-reality devices that are functional, useful, interesting, and desirable are not only technical challenges, but also artistic and practical challenges. University envi-

---

[26]Work by Jeffrey Shaw; see <http://www.zkm.de>.

[27]A vision in which processors are embedded in everyday objects and networked together would integrate computation invisibly and seamlessly into daily life. See Mark Weiser, 1991, "The Computer for the 21st Century," *Scientific American* 265(3): 94-104. Ubiquitous computing is discussed further in the section "Mobile and Ubiquitous Computing," below.

ronments are one venue where the relevant expertise and aspirations are brought together.

At Georgia Institute of Technology's Graphics, Visualization, and Usability Center, for example, computer scientist Blair MacIntyre and media theorist Jay David Bolter collaborate on the Sweet Auburn project, cross-informing augmented-reality technology and content development. They are developing applications to support tours of Atlanta's historic Auburn district, in which "ghosts" from Auburn's past appear superimposed over the landscape and address users with their stories.[28] At the MIT Media Laboratory, Hiroshi Ishii's Tangible Media group integrates art, design, and human-computer interaction to generate pre-market speculative applications such as music bottles that can be uncorked to release the music inside and "curlybots" that record and play back physical gestures.[29] A playful, speculative design approach is taken at the Computer-Related Design program at the Royal College of Art (London), home of Bishop's marble answering machine and whimsical applications ranging from a telepathic Tamagotchi[30] to a bird feeder that use principles of reinforcement learning to teach songbirds new tunes.

## COMPUTER GAMES

Computer games, having long ago left their roots as playful experiments for academic computer scientists, are emerging as a contemporary topic of computer science research because advances in many component technologies have driven burgeoning interest in "games" for serious contexts as well as entertainment.[31] Thus, for example, in the mid-1990s the Department of Defense (DOD) began to explore prospects for research collaborations among people doing modeling and simulation in defense and entertainment (including games) contexts,[32] and the Defense Advanced Research Projects Agency

---

[28]Blair MacIntyre, Marco Lohse, Jay Bolter, and Emmanuel Moreno, 2001, "Ghosts in the Machine: Integrating 2D Video Actors into a 3D AR System," pp. 80-83 in *Proceedings of 2nd International Symposium on Mixed Reality*, Yokohama, Japan, March 14-15.

[29]Hiroshi Ishii and Brygg Ullmer, 1997, "Tangible Bits: Seamless Interfaces Between People, Bits and Atoms," pp. 234-241 in *Proceedings of the 1997 Conference on Computer-Human Interaction*, ACM Press, New York.

[30]A Tamagotchi (pronounced "tom-ah-GOT-chee") is a relatively inexpensive toy containing a small liquid-crystal display, a few touch-sensitive user controls, and a program in which the image of a small creature is visible. Users can see the creature mature or, if insufficient attention is paid, see the creature die. Tamagotchi comes from the Japanese terms "tamago" meaning "egg" and "chi" as a term of endearment; it means, approximately, "lovable egg." Derived from <http://whatis.techtarget.com/definition/0,,sid9_gci213089,00.html>. Also see Anthony Dunne, 1999, *Hertzian Tales: Electronic Products, Aesthetic Experience & Critical Design*, Art Books International Ltd., London.

[31]The popular appeal and success of computer games are discussed in Chapter 2.

[32]See Computer Science and Telecommunications Board, National Research Council, 1997, *Modeling and Simulation: Linking Entertainment and Defense*, National Academy Press, Washington, D.C.

has begun to explore the potential of games for decision support. The DOD exploration gave rise to an Army-funded center at the University of Southern California, the Institute for Creative Technology; related work had already begun at the Naval Postgraduate School in Monterey, California.

Computer games offer a unique playground for serious research, not only because of the underlying allure of fun and competition, but also because important new questions arise. For example, what is a body, a surface (when infinitely malleable), or a space? How does one deal with a changing sense of time given that one can go back to a saved game? How then does one change the way one plays? How does one convey the essence of person despite screen form—gestures, and so on—varying? Most interestingly, designers of massively multiplayer online games are grappling, with a large degree of success, with the social, political, and aesthetic issues inherent in virtual worlds. What is the social contract between participants, and between the participant and the designer? What are the consequences of conflicts in the virtual world, and to what degree should those consequences be determined by the online population, versus the administration? How should people deal with the distribution of authorship in an environment where narratives are participatory and emergent? How does one foster organic, self-organizing social structures in a virtual world? How does a designer make places people want not only to visit, but also to inhabit for hundreds or thousands of hours over the course of several years? These questions raise various issues for a number of computer science fields, including information retrieval, database management, and computer graphics, to name a few—though such questions are not purely CS ones, but rather questions that are truly transdisciplinary. There is evidence that CS is beginning to address some of these questions (e.g., see the special issue "Game Engines in Scientific Research" in the *Communications of the ACM*, January 2002).[33]

## NARRATIVE INTELLIGENCE

In the early 1990s, a group of graduate students at the MIT Media Lab formed a new reading group, which they called narrative intelligence (NI).[34] The group explored issues at the intersection of narrative and both human intelligence and AI, seeking to develop a dialogue between new computational concepts and technologies and the insights of literary theories such as poststructuralism and semiotics. The group came together with an understanding of, and the desire to

---

[33]One reviewer observed that the CS research agenda was being only modestly influenced as of November 2002.

[34]Marc Davis and Michael Travers, 1999, "A Brief Overview of the Narrative Intelligence Reading Group," pp. 11-16 in *Proceedings of the 1999 AAAI Symposium on Narrative Intelligence*, Michael Mateas and Phoebe Sengers, eds., AAAI Press, Menlo Park, Calif.

reconcile, the contradictions and incompatibilities between these two world views: AI technology focused by and large on formal, logical representation and objectivity, whereas the analytical tools provided by new literary theories focused on subjectivity, multiplicity, and the limitations of formalism. The pragmatics of negotiating the differences between these world views led to a creative foment. The group flourished, exploring issues in the philosophy of mind, media theory, HCI, psychology, social computing, constructionism, and AI, developing theories and applications in all these areas, influencing the direction of the doctoral program at the Media Lab, and connecting to a wider network of researchers who joined in the group's discussions over e-mail. Narrative intelligence as a field was born.

NI research obviously incorporates influences from a variety of fields. Artificial intelligence, with tools to model human emotion, personality, and narrative abilities, provides a framework from which much of the research grows. Psychology, especially narrative psychology, generates explanations of the human ability to understand the world through narrative, creating a basis for systems that model or support this ability. Art research raises new questions about the nature of narrative representation, keeping the concept of narrative fresh. Cultural studies analyze hidden cultural narratives, including the stories AI researchers tell through their research. Literary studies examine the nature of narrative in traditional and interactive forms. Drama provides understanding of the real-time performance of narrative. This emphasis on mixing technology development with artistic and humanistic perspectives is unusual in AI. It has supported the generation of new research fields within AI, such as lifelike interactive computer characters, as well as an increase in cross-disciplinary engagement between AI and other fields.[35]

At the same time, narrative trends took on importance in related fields. The concept of supporting human narrative understanding through the interface of human and computer began to gain ground in the field of HCI. Work in media studies on hypertext and interactive fiction was inspiring a generation of systems that support narrative in new ways. Within AI, this interest began to spur research in AI for interactive fiction and entertainment,[36] including interactive computer

---

[35]Narrative as a topic of research is not new to AI. In the 1970s and early 1980s, there was substantial interest in modeling story understanding and story generation, particularly by Roger Schank's research group at Yale University. Programs developed by the group—which were able to generate stories and answer questions about them, albeit in limited ways and domains—illustrated theories of human understanding and the structure of knowledge in the mind. Massive, unwieldy, and hard to extend, these systems ran into trouble during the "AI winter" of the 1980s. Researchers, seeking to combat the image of AI as never living up to its inflated claims, favored clearly defined problems with easily measurable outcomes—a situation that is not conducive to this kind of creative work—and, therefore, narrative fell out of favor for a decade as a research topic in AI.

[36]Joseph Bates, 1992, "Virtual Reality, Art, and Entertainment," *Presence: The Journal of Teleoperators and Virtual Environments* 1(1): 133-138.

**BOX 4.2**
**Virtual Characters—Improvisational Actors**

The research, development, and commercialization of a new class of intelligent-agent applications that reflect a philosophy of learning through play is the focus of Extempo Systems. Software-driven virtual characters, which Extempo Systems' founder Barbara Hayes-Roth[1] calls improvisational actors, can take the form of a human, an animal, or just about anything the imagination can dream up, to interact directly with people visiting a Web site. What makes them different from other types of intelligent agents is their ability to improvise stories. According to Hayes-Roth, this design allows the actors to achieve their goals while providing engaging conversation for many different Web site visitors. One of Extempo's creations, a dog named Jack, seeks to achieve six goals when people interact with him: get them involved, guide them to target information, gather customer data, personalize the experience, delight them, and build site loyalty. Jack accomplishes his goals by improvising appropriately in response to a situation. For example, as a visitor enters a site, Jack says what he can do for the person. If the visitor stops "talking," Jack is programmed with a few things to say based on the content of the conversation thus far. Like a human, Jack is designed to have a certain protocol involving turn taking and interruption. This allows him to have characteristics of intelligent conversation, such as making functional transformations and analogizing between topics.

---

[1]See <http://www.extempo.com> for additional information about Extempo Systems and the work of Barbara Hayes-Roth, who was a briefer at the committee's January 2001 meeting held at Stanford University.

characters and interactive plots. Many of these research areas explicitly draw on the arts and drama as a source of inspiration. With the growth of the computer game industry has come an interest in new game forms that support narrative in more complex and interesting ways than a stereotypical shoot-and-kill form.

Research in NI is flourishing, with applications in a variety of areas. Narrative interfaces explore possibilities for making interfaces more usable by incorporating elements of story, for example by embodying interaction in a storytelling character. Artificial agents can themselves be designed to use narrative, as humans do, to make sense of the world and each other (see Box 4.2). Researchers are developing systems to support human storytelling, as in the case of plush toys that children can program to tell their stories to families and friends.[37] Databases of stories allow people to search for and share stories pertinent to their experiences.[38] Stories can be automatically generated, perhaps in response to input from human users. Interactive digital video allows video sequences to be generated interactively, telling

---

[37]Marina Umaschi, 1997, "Soft Toys with Computer Hearts: Building Personal Storytelling Environments," pp. 20-21 in *CHI '97 Proceedings*, ACM Press, New York.

[38]See Justin Cassell and Jennifer Smith, 1999, "The Victorian Laptop," pp. 72-78 in *Narrative Intelligence: Papers from the 1999 Fall Symposium*, Technical Report FS-99-01, Michael Mateas and Phoebe Sengers, eds., AAAI Press, Menlo Park, Calif.

interactive stories.[39]   The field of interactive fiction and drama has exploded,[40] including the subfield of interactive computer characters, or characters with emotion and personality who respond to human users in the context of a story.[41]   A complementary area of narrative intelligence studies the stories that AI researchers themselves tell about what they are doing.[42]   Sometimes, analysis of these stories can lead to new forms of AI technology by building on alternative stories.[43]

In this explosion of research, the interdisciplinary engagement begun by the NI group at the Media Lab remains present—in work taking place in traditional computer science departments, in cross-disciplinary arenas like the Media Lab, in humanities and arts departments that incorporate new media such as Georgia Institute of Technology's School of Literature, Communication, and Culture,[44] and in the computer game industry.

## NON-UTILITARIAN EVALUATION

As discussed above, artists traditionally use evaluation techniques that differ radically from those of computer scientists, with little interest in formal user studies and more interest in social impact, cultural meaning, and the potential political implications of a technology. They seek to provoke as well as to understand the user. There is an opportunity to develop hybrid evaluation methodologies to combine the broader concerns of artists with the narrower and more structured methodologies of HCI. For example, Angela Garabet, Steve Mann, and James Fung use strategies that are open-ended and interpretive[45] to evaluate users' reactions to wearable computing designs. Interestingly, they demonstrate that users are more open to and accepting of new technology that is presented as the product of a commercial venture rather than as art. Jonas Lundberg and colleagues uninten-

---

[39]See Glorianna Davenport and Michael Murtaugh, 1997, "Autonomist Storyteller Systems and the Shifting Sands of Story," *IBM Systems Journal* 46(3): 446-456.

[40]For example, see Peter Weyhrauch, 1997, *Guiding Interactive Drama*, Ph.D. Thesis, School of Computer Science, Carnegie Mellon University, Technical Report CMU-CS-97-109, Pittsburgh, Pa.

[41]See Bruce Mitchell Blumberg, 1996, *Old Tricks, New Dogs: Ethology and Interactive Creatures*, Ph.D. Thesis, MIT Media Laboratory, Cambridge, Mass.

[42]See N. Katherine Hayles, 1999, *How We Became Posthuman: Virtual Bodies in Cybernetics, Literature, and Informatics*, University of Chicago Press, Chicago.

[43]See Philip E. Agre, 1997, *Computation and Human Experience*, Cambridge University Press, Cambridge, U.K.; also Phoebe Sengers, 1998, *Anti-boxology: Agent Design in Cultural Context*, Ph.D. Thesis, School of Computer Science, Carnegie Mellon University, Technical Report CMU-CS-98-151, Pittsburgh, Pa.

[44]Described in Chapter 6.

[45]For example, because evaluators are trying to understand how people react to a system, users might not be told about the purpose or operation of a system. See Angela Garabet, Steve Mann, and James Fung, 2002, "Exploring Design Through Wearable Computing Art(ifacts)," *Computer-Human Interaction 2002, Interactive Poster: Fun*, pp. 634-635.

tionally achieved similar results in their explorations of a provocative technology, a refrigerator that videotaped its users, ostensibly allowing those who shared the refrigerator to find out if someone had stolen their food. Although the goal was to confront users with negative aspects of technology, users who saw the "product" demonstrated in an ostensibly commercial presentation were surprisingly enthusiastic.[46] Such results may motivate some artists to be more interested in collaborations with commercial objectives than they might otherwise be.

Evaluation techniques drawing on both HCI and arts traditions could rigorously examine not only the usability and utility of software and electronic products, but also the meanings they may take on in users' everyday lives, the background cultural assumptions that underlie them (for example, the assumptions designers make about what users are like), and their potential impact on current cultural issues and debates, such as intellectual property issues. At the same time, standard HCI techniques appropriately adapted to the goals of artists (often far removed from issues of usefulness and efficiency that current techniques can address) may help improve the sometimes opaque design of interactive artwork. To achieve these goals will likely require a fundamental rethinking of the notion of user tests, as well as other evaluations. In an early example of what such work might look like, artist-designers Anthony Dunne and Fiona Raby evaluated the Placebo project, electronically enhanced furniture that makes users aware of activity in the electromagnetic spectrum, through open-ended interviews with users combined with photographic portraits of users with their devices.[47] Such techniques allow designers to do evaluation in a form that is to some extent recognizable and understandable to HCI practitioners, while exploring issues that matter to artists, such as the subjective nature of user experience, the stories that give devices not only functionality but also meaning in human context, and the messages that information technologies intentionally or unintentionally communicate to users. There is a need to develop a repertoire of evaluation techniques appropriate for these more open-ended questions that is as wide and deep as that already available for the relatively well defined problems of usability and efficiency.

## EXPERIMENTAL CONSUMER PRODUCT DESIGN

Computing is creating new challenges as it moves into everyday life. The impact of IT on everyday culture is felt particularly strongly through electronic consumer products such as handheld computers,

---

[46]Jonas Lundberg, Aseel Ibrahim, David Jönsson, Sinna Lindquist, and Pernilla Qvarfordt, 2002, "The Snatcher Catcher: An Interactive Refrigerator," Short Paper, *NordiChi 2002*, pp. 209-211.

[47]Anthony Dunne and Fiona Raby, 2001, *Design Noir: The Secret Life of Electronic Objects*, August/Birkhaeuser, Basel, Switzerland.

music storage and playback devices, and electronic toys. Designers Fiona Raby and Anthony Dunne have argued that such consumer products are currently designed much the same way as Hollywood movies: They are generally uncontroversial, focused on socially acceptable needs, and broadly marketed, and they serve an optimistically idealized lifestyle. In the analogy to film, they note that the alternatives to mainstream Hollywood film—such as film noir, experimental film, and independent cinema—consistently develop techniques of narrative and visual style that are later adopted by Hollywood, thus in effect serving an R&D role.[48] Likewise, experimental designers often develop ideas that, while often not immediately marketed, influence and eventually help redirect contemporary design practices in marketed products.

Experimental designers explore a range of issues and ideas that are often different from those of individuals working in specific product fields who are more constrained by the demands of the market. Their work often explores issues at the intersection of product design and social issues. For example, the 2002 show of the Interaction Design program at the Royal College of Art included Pedro Sepulveda's architectural designs responding to fears and anxieties about cell phone radiation. There is tremendous potential for ITCP not only in reconfiguring existing consumer electronic applications, but also in imagining and building prototypes for new applications and markets.

## MOBILE AND UBIQUITOUS COMPUTING

As hardware components become smaller, faster, and cheaper, IT is being embedded into more and more physical devices, linked together through (often wireless) networks. Networked systems of embedded computers (EmNets) will be largely invisible but extremely powerful, allowing information to be collected, shared, and processed in new ways. EmNets promise significant changes in environmental and personal monitoring as well as scientific research. Thousands or millions of sensors could monitor the environment, the battlefield, the home, the office, or the factory floor; smart space containing intelligent surfaces and appliances would provide access to computational resources.[49]

However, product development often follows the path of least resistance, resulting in products that are technically new but do not take full advantage of the broad conceptual design space that is opened up by mobile and ubiquitous technologies. For instance, it is not uncommon to simply replace discrete analog or digital electronics with an inexpensive microcontroller, or to imagine faster or smaller

---

[48]Dunne and Raby, 2001, *Design Noir*.

[49]This paragraph is drawn from Computer Science and Telecommunications Board, National Research Council, 2001, *Embedded, Everywhere: A Research Agenda for Networked Systems of Embedded Computers*, National Academy Press, Washington, D.C.

versions of existing applications. These approaches reflect an internal technical logic and a safe approach to introducing new products to consumers in a competitive marketplace. As a result, they can ignore some of the larger challenges in designing devices that could and should meaningfully support everyday quality of life.

The views of social scientists, designers, and artists are needed to address the potential implications of EmNets. Imagine living in a world where it is impossible for anyone to get lost anywhere. Cars, wireless devices, laptops, TVs, and even articles of clothing will always be able to tell you and others exactly where you are. Imagine also living in a world where devices recognize a person's speech, emotions (through physical expressions and tone of voice), and likes and dislikes, and respond accordingly. Now, many who use e-mail and the Web enjoy relationships that involve real, but digitized, humans. Will technology make so much available without complaint that the need for interpersonal relations dissolves? Will that relationship with technology generate an illusion of satisfaction, knowledge, creativity, and life? These are the kinds of questions that influence IT research, and they are also the kinds of questions that inspire artistic exploration via ITCP[50]—both activities where government-funded academic research or philanthropically supported arts-based activity can explore ideas not likely to flow from conventional commercial efforts.[51]

The development of smart appliances for the home provides an interesting case study. Many of the gadgets being developed today— from refrigerators that can determine when to order more milk and scales that monitor users' weight gain and suggest low-calorie recipes to home entertainment systems that remember the preferred settings of different users in the home—are obvious extensions of already-existing technologies. The approach of the information arts, in which technical questions are seen as interrelated with social and cultural questions, lends itself well to a more fundamental shift in the design of smart appliances to support not only new technologies but also new, better, and/or more interesting ways of living in the home. The Domestic Environments project of the Equator research collaboration,[52] funded by the U.K. Engineering and Physical Sciences Research Council, is one model of how computer scientists collaborating with artist-designers and social scientists can develop appliances that are interesting both technically and socially. One of the appliances is a "drift

---

[50]Although this chapter focuses on how ITCP, or the information arts, might influence the discipline of computer science, clearly ITCP can (and does) influence the agenda for artists and designers as well. In this sense, IT does much more than provide tools to the arts and design disciplines.

[51]As a comparison, consider the challenge of developing new kinds of content and applications for broadband networks. As CSTB has observed previously, there is an important role for academic research. See Computer Science and Telecommunications Board, National Research Council, 2002, *Broadband: Bringing Home the Bits*, National Academy Press, Washington, D.C.

[52]See <http://www.equator.ac.uk/>.

table" designed to promote daydreaming and reflection. A "window" built into the table shows images of the British countryside. As objects are placed on the table, the window begins to drift slowly over the countryside.

The focus on security resulting from the September 11, 2001, terrorist attacks may well be a theme worthy of exploration. As technologists push ahead an agenda for more robust computer security, the information arts would promote an agenda of how to improve security with the least harm to society (or even allowing for the possibility of improved security and a net positive gain to quality of life).[53] Additional areas of interest include work inspired by the implications of intellectual property law and policy for the digital environment (see Chapter 7 for a discussion of digital copyright) and bioinformatics.

• • • • • • • • • • • • • • • • • • • • • • • • • • • • • • • • • • • • • • • • •

## CONCLUSION

Two different kinds of intersection between IT and creative practices are presented in Chapters 3 and 4. On the one hand, Chapter 3 looks at the use of information technology as a medium for art and design practices, suggesting that computer science can support ITCP. On the other, Chapter 4 looks at ways in which art practice and design and computer science can become fused, leading to new fundamental insights into the nature of computer science itself. In practice, these two kinds of intersection are not disjoint. For example, imagine the development of new word processors with input from professional writers (see Box 4.3).

In policy circles, a vigorous debate has been taking place in recent years about whether knowledge production at large is shifting from discipline-bound, strongly bounded, and relatively stable models to transdisciplinary, loosely coupled, and transient ones.[54] There is little to be gained by preferring either multidisciplinary or transdisciplinary exchanges; both have their place and are capable of generating useful and interesting results.

> There is little to be gained by preferring either multidisciplinary or transdisciplinary exchanges; both are capable of generating useful and interesting results.

---

[53]Of course, new security sensitivities could have the side effect of complicating ITCP activities that seem to take advantage of weaknesses in order to manipulate conventional IT.

[54]David et al. (1999) attacks Gibbons et al. (1994); Gibbons et al. (2001) responds. Paul David, Dominique Foray, and W. Edward Steinmueller, 1999, "The Research Network and the New Economics of Science: From Metaphors to Organizational Behaviors," in *The Organization of Economic Innovation in Europe*, A. Gambardella and F. Malerba, eds., Cambridge University Press, Cambridge, U.K., and New York; Michael Gibbons, Camille Limoges, H. Nowotny, S. Schwartzman, P. Scott, and M. Trow, 1994, *The New Production of Knowledge. The Dynamics of Science and Research in Contemporary Societies*, Sage, London; Michael Gibbons, Helga Nowotny, and Peter Scott, 2001, *Rethinking Science: Knowledge and the Public in an Age of Uncertainty*, Blackwell, Malden, Mass.

## BOX 4.3
## Information Technology and Word Processing

Consider IT tools that support writing, such as the word processor. Most word processors are based on an underlying representation of text as a long string of characters, each having some properties affecting its surface appearance. This was not a deeply thought out metaphor for the process of writing; rather it evolved through generations of automation tools—the word processor as a better typewriter, or a software tool for automating typesetting. Some word processors evolved to have a rudimentary structural representation of writing—outlines, for example—but the basic representation shines through even style sheets and wizards. An apparent benefit of this approach is that it supports all kinds of writing equally well: IT does not attempt to reinforce one writing style or approach to composition in favor of another (ignoring, for the moment, the effects of grammar checkers and other add-ons that attempt to critique the actual writing). However, there is little doubt that the ease with which revisions can be made, ranging from typographical corrections to particular large-scale structural changes, has influenced the approach one takes to writing. This design reflects choices by software developers to structure writing as a simple string of characters, provide context-blind editing operations such as copy and paste, and let users work through the higher-level concepts and details for themselves.

As an example of a realization of a completely different set of design choices, consider Dramatica Pro— a popular tool for developing screenplays.[1] Superficially, it is still a tool for supporting writing. However, it uses a domain model for screenwriting, including characters, themes, scenes, plot progression, character development, and so on. It claims 32,768 "story forms," allows four different viewpoints on the story being told ("throughlines"), and contains models of how plots develop, how character traits relate to one another, and how characters can respond to challenges. An intended side effect of this approach is that it forces the user to consider certain aspects of structure, character, plot, and theme, claiming the benefits of "expanding your creative potential." Without passing judgment on this particular tool, it is fair to note that its design reflects different choices—an attempt to deeply embody domain knowledge, a focus on supporting higher-level details—and, in fact, provides very limited editing operations such as copy and paste and virtually ignores issues related to the surface appearance of text. It is intended to address a much narrower set of tasks than a general-purpose word processing tool and, as such, would be of little use in writing a report such as this one. Could a writer create a new kind of screenplay for a radically new kind of film using this tool, or would such a creation be outside its scope? Is it conceivable that a writer could push against the boundaries created by such a tool, abuse its constructs in various ways, and produce something surprising to the tools designers? Does having these boundaries against which to push perhaps expand creative potential even more than having structures within which to work?[2]

Not enough is known about the deep impact of such design choices on users. However, it could prove very useful to explore, as an example, different kinds of word processors that could be designed, for instance, based on radically different notions of what writing is about, or on some deep criticism of the underlying assumptions embodied in typical word processors from the point of view of writers. The result could be a meaningful intellectual shift in computer science, as well as a tool that is much more useful for writers— and possibly for the rest of us as well.

---

[1]This discussion is based on version 4.0 of Dramatica. See <http://www.dramatica.com/> for details.

[2]Dramatica is intended as a tool for developing story lines, not for producing the screenplay document. However, it does interface with a more traditionally styled word processor, tailored to the unique needs of producing screenplays. This is another example of the kind of choices made by a designer. Further research into innovative tools for supporting creative writing might make it possible to integrate both metaphors within a single tool.

But both the multidisciplinary and the transdisciplinary models make clear that there is a continuing need to maintain the integrity of the traditional disciplines, both in the arts and in the sciences. Without a disciplinary frame, the richness of disciplinary practices, methodologies, and concepts can become lost, leaving an oversimplified cross-disciplinary knowledge domain. This danger exists when any practice is digitized in the absence of an appropriate model, as for example in arts education when young people have become wedded to the pre-scripted options of packaged applications and are only capable of creating PhotoShop art. What Paul David and his co-authors fear would become "cut-price research motels" in scientific research[55] corresponds closely to the degeneration of artistic quality that is possible where electronic art forms (or media art or the modish "new media") have been cut loose from their deep connections to older and richer art practices.

Finally, it should be noted that the transient, loose coupling of transdisciplinary creativity runs an ever-present risk of premature bureaucratization.[56] A single successful outcome is a necessary but by no means sufficient reason to continue cross-disciplinary work in the same vein. In some cases, the outcome of a rich experimental device is best evaluated and further developed in the separate but transformed disciplines that contributed to it. In other cases, however, the committee has found persuasive evidence of the need for the sustained bridging of disciplines, involving the development of both individual practices and a community of researchers in the cross-disciplinary area with correspondingly innovative institutional structures (and these are discussed in Chapter 5).

---

[55]David et al., 1999, "The Research Network and the New Economics of Science," p. 334.

[56]As one reviewer notes, there is much to learn from history. See Carolyn Marvin, 1998, *When Old Technologies Were New*, Oxford University Press; Brian Winston, 1998, *Media Technology and Society*, Routledge, London; and Paul N. Edwards, 1996, *The Closed World: Computers and the Politics of Discourse in Cold War America*, MIT Press, Cambridge, Mass. But much of ITCP history is undocumented or otherwise not very accessible, and therefore, it is unteachable in an organized way. As a result, practitioners are unaware of precedents and relevant prior research, and they constantly reinvent the wheel.

# 5 Venues for Information Technology and Creative Practices

ndividuals and groups involved with information technology and creative practices (ITCP) benefit from participating in venues that support, motivate, and display this type of work. Such venues may occupy physical or virtual spaces, vary widely in scale and scope, range from loosely organized collectives to formal programs, and be either free-standing or connected to established institutions. They can offer important benefits such as access to tools, information resources, work spaces, funding,[1] and opportunities for communication among practitioners, those who fund and display ITCP work, and audiences. This chapter describes and analyzes the evolution and characteristics of these venues for the purpose of providing guidance for the design of future ones.

The first section provides a historical perspective on studio-laboratories, which bring together different domains of knowledge, research, and practice—thus combining the artist's (or designer's) studio with the scientist's (or engineer's) laboratory. This section also introduces the three classes of modern studio-laboratories: multifaceted new-media art and design[2] organizations (typically non-profit), mechanisms for public display, and applied research activities in corporations. The remainder of the chapter discusses examples of these types of organizations and activities and draws distinctions between patterns in the United States and those abroad. Programs associated with academic institutions are an important special case and are discussed in Chapter 6.

---

[1]Funding issues are discussed primarily in Chapter 8.

[2]New-media art (and design) can be loosely characterized as art (and design) that uses IT in a significant way. Internet art (Net art) is a subset of new-media art, because the Internet is a subset of IT. Digital art is also a subset of new-media art, because new media could incorporate non-digital technologies or content.

• • • • • • • • • • • • • • • • • • • • • • • • • • • • • • • • • • • • • ◀

# STUDIO-LABORATORIES

## HISTORICAL PERSPECTIVE

In the evolution of creative work, the role of individuals in creating new practices and of disciplines in providing tools and methodologies to be reworked is well understood. But the essential role that institutions play in making it possible for individuals and disciplines to come into contact and flourish can be more difficult to recognize. An accurate history of 20th-century developments related to ITCP would counter the widespread but false impression that there has been a renaissance of creativity enabled uniquely by the computer, and would make clear that a gradual development of new institutions, especially the studio-laboratory, has also played a central role.

The term "renaissance," though, may in fact be appropriate, because the role of institutions in the last century parallels the central, often unrecognized role that they played in the Renaissance. In 1950, art historian Erwin Panofsky pointed out that the age of Leonardo and Durer was one of turbulence, decompartmentalization, and in particular great social mobility after the cloistered separation of the Medieval worlds of theory and practice. The great advances, he added, were indeed made by the instrument makers, artists, and engineers, not the professors. But what made this possible were new institutions, like the academies, that served as "transmission belts" between previously separated domains of knowledge and practice. Perhaps blinded by the brilliance of the rare figure of the da Vincian creator, one may lose sight of the importance of the dilettanti who frequented the academies—people who were "interested in many things."[3] Today's studio-laboratories are similarly populated.

If the current era is, like Panofsky's Renaissance, a period of hybridity and interdisciplinarity, then it can also be described as a period where, in many different, local places, experts of diverse kinds, interested in "many things," meet and learn from one another. The late 20th century has generally been described as a period of post-modernism, in which the stable values of the past were no longer pertinent and were being replaced by a medley of ideas from different sources. Instead of taking the general view that sees ideas as hopelessly jumbled together, one can look for the complex dynamics that arise when specific ideas come into contact with each other at a variety of local, concrete places. This perspective corresponds with recent trends in social theory, in which there is less focus on the individual

An accurate

history of the 20th

century would

counter the

widespread but

false impression

that there has

been a

renaissance of

creativity enabled

uniquely by the

computer.

---

[3] Erwin Panofsky, 1952, "Artist, Scientist, Genius. Notes on the Renaissance Dammerung," *The Renaissance: A Symposium*, Metropolitan Museum of Art, New York.

units of society and more on the complexity that results when they interact with each other. This allows us to move from thinking about art/design and science/technology as two distinct, separate spheres, and to look instead at how they interact with each other.[4] From this perspective, the studio-laboratory can be viewed as a hybrid institution where such interaction can occur. A useful demarcation point can be found in the famous Bauhaus slogan "art and technology, a new unity," in the early 1920s. It is here that a cursory genealogy of the studio-laboratory can begin. See Box 5.1.

## THREE CLASSES OF MODERN STUDIO-LABORATORIES

There are three classes (or phases) in the development of modern studio-laboratories: art/design-driven technology development, public diffusion and critical debate, and industrially sponsored applied research. Taken together, the three classes theoretically form a continuum involving upstream experimentation, public diffusion of results, and downstream development. These classes can easily be merged into a single concern for more innovation and creativity in the art-design-technology-science intersection. However, the reality is that the three classes came into being at different times and are rarely in close productive cooperation. Thus, it is not surprising that some aspects of the continuum have been more successful than others, leaving an overall impression of unevenness and some significant gaps.

Figure 5.1 conveys in broad outline the increasing frequency with which studio-laboratories were founded, especially after the 1960s. In the first phase, research and production were oriented principally toward the creation of singular, visionary artworks using technology and little concerned with, or constrained by, wider application (or usability) outside the immediate aesthetic context. In the second phase, toward the end of the 1970s, specialized institutions were planned and established to focus on the presentation of new technological art in public arenas, and critical analysis and recognition of such work. The third phase, beginning in the 1980s alongside the opening up of mass markets for personal computers, packaged end-user applications, and cultural commodities like musical instruments and video games, was oriented toward applied research and development.

Among the most oft-cited examples from the first phase was Experiments in Art and Technology (E.A.T.), founded by artist Robert

[4]Karen Knorr-Cetina, 1999, *Epistemic Cultures: How the Sciences Make Knowledge,* Harvard University Press, Cambridge, Mass. "I also reject the notion that there is a sharp distinction between technical (instrumental, productive, rational) activities and symbolic processes. It is the exclusionary definition of the two that poses the problem."

## BOX 5.1
## The Bauhaus

Influenced by the 19th-century British arts and crafts movement, the Bauhaus is commonly thought of primarily in terms of the Modern design style with which it is usually associated, or in terms of its formative influence in design education. But aside from its aesthetics, one of the basic issues the Bauhaus worked on was the tension between commitments to research on basic artistic principles and to making saleable, machine-reproducible objects. This tension dogged it through its brief but highly influential 14-year existence. Constantly in turmoil, from its founding in the disastrous wake of World War I to suppression by the Nazis, the Bauhaus was never a "stable, communal, harmonious ensemble."[1]

An important example of the research made possible in the Bauhaus context was the work of the "pure" artists who were employed to bring "exact experimental" methods to the realm of art.[2] The painter Paul Klee, for instance, created a substantial body of theoretical and pedagogical work at the Bauhaus, much of it concerned with the problem of visual notation of dynamic form. This research-production was to a large degree out of sync with the immediate problems of mass production and housing. However, Klee's sustained investigation of visual form in terms of dynamics proved to have enormous influence on other fields outside art, such as music composition and film animation, and on computer graphics decades later. All this work was carried out using traditional means of drawing and painting, and was concerned not with directly representing the Machine Age of the Bauhaus, but, more deeply, with the logical primitives needed for visually representing dynamic relationships. Klee's work was deeply reflective of broader cultural issues of its time, including the intoxication with speed, transportation, and new models of time. But it was also beyond its time, in that it opened up a rich vein of "exact intuitions" (as Klee himself might have put it) whose implications were directly applicable to later practitioners working with newer technologies.

The Bauhaus ideal of an "experimental mini-cosmos" mutually shaped by art, engineering, science, and philosophical concepts[3] served as inspiration and template for subsequent organizations crossing the boundaries between technology and culture. This occurred partly through people, as many of the Bauhaus leaders immigrated to the United States. But, more generally, the idea of an institutional site for research to which master artists would attach themselves, if provisionally, has been influential, continuously invoked in the foundation of subsequent art/design and technology research centers since the 1960s. There were other important influences from the early 20th-century avant-garde on more recent art-technology movements. But the Bauhaus developed what sociologist Henri Lefebvre has called a "specific rationality," which was new to the cultural sphere.[4] This specific rationality—characteristic of the studio-laboratory—was inclusive, refusing a sharp distinction between the techno-scientific and the symbolic-expressive realms. For the modernist master composer Pierre Boulez in the 1970s, or the architectural theorist Heinrich Klotz in the 1980s, or computer scientist Pelle Ehn in the 1990s, the Bauhaus model inspired a holistic approach to co-development of new art and new technologies, both informed by theoretical research.[5]

[1]Elaine S. Hochman, 1997, *Bauhaus: Crucible of Modernism,* Fromm International, New York.

[2]Paul Klee, 1928, "Exact Experiments in the Realm of Art," *Paul Klee: The Thinking Eye, Notebooks of Paul Klee,* Jürg Spiller, ed., Lund Humphries, London.

[3]Peter Galison, 1990, "Aufbau/Bauhaus: Logical Positivism and Architectural Modernism," *Critical Inquiry* 16 (Summer): 709-752.

[4]Henri Lefebvre, 1991 [1974], *The Production of Space,* Blackwell, Oxford.

[5]Heinrich Klotz, 1990, *The Center for Art and Media Technology Karlsruhe. An Architecture Competition,* Okotogon, Stuttgart-München; Pierre Boulez, 1985, *Le modèle du Bauhaus. Points de repère,* Editions Seuil, Paris.

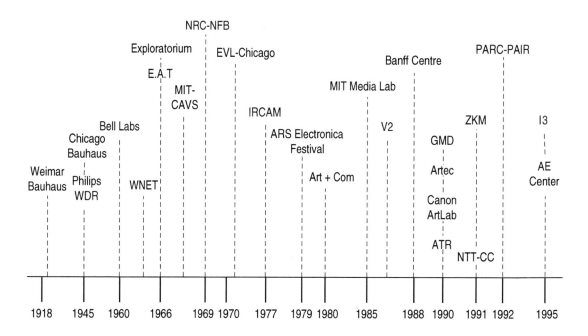

FIGURE 5.1 Studio-laboratories in the 20th century.

AE Center—Ars Electronica, Austria
ATR—ATR Media Integration and Communications Systems Laboratories, Japan
E.A.T.— Experiments in Art and Technology
EVL–Chicago—Electronic Visualization Laboratory, University of Illinois, Chicago
GMD—German Research Center for Information Technology
I3 networks—Intelligent Information Interfaces, program of European Commission Fifth Framework for Research
IRCAM—Institut de Recherche et Coordination en Acoustique et Musique (Institute of Research and Coordination in Acoustics and Music), Paris
MIT–CAVS—Massachusetts Institute of Technology–Center for Advanced Visual Studies
NRC–NFB—National Research Council Canada–National Film Board
NTT–ICC—Nippon Telegraph and Telephone InterCommunication Center, Tokyo
PARC–PAIR—Xerox Palo Alto Research Center–Artist in Residence Program
V2—V2_Organisation, Institute for the Unstable Media, Rotterdam
WDR—West German Broadcasting Network
WNET—WNET Television Workshop Artist-in-Residence Program
ZKM—Zentrum für Kunst und Medien (Center for Art and Media), Karlsruhe, Germany

Rauschenberg and Bell Labs physicist Billy Klüver in New York in 1966. The goal of E.A.T. was to establish "an international network of experimental services and activities designed to catalyze the physical, economic and social conditions necessary for cooperation between artists, engineers and scientists." The research role of the contemporary artist was understood by E.A.T. as providing "a unique source of experimentation and exploration for developing human environments of the future."[5] At the same time, other Bell Labs scientists were also engaged in collaborative research, in computer graphics and vision, music, and acoustics.[6] Also during the late 1960s, at MIT, the Hungarian artist and Bauhaus affiliate Gyorgy Kepes founded the Center for Advanced Visual Studies, providing a stable location for collaboration between artists in residence and university-based scientists and engineers. In the second phase, the earliest self-standing art-technology centers appeared—notably the Institut de Recherche et Coordination en Acoustique et Musique (IRCAM) in Paris—with a focus on the artistic rather than the industrial potential of information technology (IT) and electronics. Composer Pierre Boulez launched the IRCAM based on a conception of research/invention as the central activity of contemporary musical creation; Boulez invoked the "model of the Bauhaus" as cross-disciplinary inspiration for what he considered the necessary and inevitable collaboration between musicians and scientists.[7] The modernist autonomy of the computer music research at IRCAM did not prevent it from also orienting its program toward a wider cultural public, aiming to put an otherwise forbidding musical language in the context of the history of 20th-century music and intellectual life.

The second phase, which incorporated festivals, exhibitions, commissions, and competitions of electronic art, marked an increased commitment of both public administrations and private corporations toward exposing the most radical media-based creativity to a wider public. As festivals such as Ars Electronica became global in scope beginning in the 1980s, so also plans were drawn up in most advanced industrial countries to establish permanent centers able to incorporate a dual research/development and public education mandate. Among the most conspicuous were the Zentrum für Kunst und Medien (ZKM) in Germany and the NTT InterCommunication Center in Japan.

In the United States, there has been no investment at a comparable scale in active commissioning, exhibition, and public programming of ITCP. Some smaller efforts have been made at a consistent level of quality and commitment, notably in science museums, such as the Exploratorium in San Francisco. New-media artists and techno-art engineers speak with a common voice in lamenting the lack of well-developed circuits for public exhibition, review, and systematic documentation. This was often expressed to the committee as the absence of European-style institutions in the United States that combine re-

---

[5]Experiments in Art and Technology (E.A.T.), 1969, *Experiments in Art and Technology Proceedings*, E.A.T., New York.

[6]Bell-Telephone, 1967, "Art and Science: Two Worlds Merge," *Bell Telephone Magazine*, November/December. Also see Box 6.3 in Chapter 6.

[7]Pierre Boulez, 1985, *Le modèle du Bauhaus. Points de repère*, Editions Seuil, Paris. In terms of computer expertise, IRCAM depended heavily on the U.S. academic computer music community, initially duplicating the computer system and software in use at Stanford University's Center for Computer Research in Music and Acoustics.

There is an

absence of

European-style

institutions in the

United States that

combine research

in contemporary

media arts with

large collections

and public

displays.

search in contemporary media arts with large collections and public displays. Artists worry about a lack of contact with the American public, and, more specifically, seek intelligent and informed responses to created installations and wider professional and disciplinary associations.

Not that ZKM and other such centers are without critics. The German philosopher and critic Florian Roetzer analyzed the media-center bandwagon of the late 1980s, when he commented sardonically that "everywhere there are plans to inaugurate media centres, in order not to lose the technological 'connection.' . . . This new attention is supported by the diffuse intention to get on with 'it' now, the contents remaining rather arbitrary, so long as art, technology and science are somehow joined in some more or less apparent affiliation with business and commerce."[8] Roetzer was then not alone among critical intellectuals in harboring a deep ambivalence about these institutional developments, fearing that they would serve only to accelerate the public acceptance of automation in everyday life, on the one hand, and to co-opt artists—"with their purported creativity"—into becoming commercial application designers, on the other. As it turned out, explicitly designed linkages between art, research, and industrial innovation developed a good deal beyond Roetzer's cynical prognostications, and rather quickly became the basis for the third phase of the contemporary studio-laboratory.

The third phase was marked by the establishment of the MIT Media Laboratory. Many observers would probably count the MIT Media Lab as the most successful, as well as widely imitated, model for industrially sponsored research on art, design, and new-media technologies. (Because of its connection to an academic institution, the lab is discussed in Chapter 6.) The Media Lab's large-scale, precompetitive industrial consortia are global in scope and have inspired a number of institutional responses in other countries. For instance, the Swedish initiative of a network of six interactive institutes, loosely interacting but also differentiated according to technology and artistic focus, grew in large measure out of a national initiative to attract and retain a critical mass of researchers while still decentralizing the activity to both large and small cities.[9]

During the third phase, a number of technology companies attempted to harness artistic expertise to further corporate goals. For example, Interval Research Corporation assembled an unusually diverse research staff, including artists and filmmakers as well as computer and social scientists. Throughout the 1990s, Xerox Palo Alto Research Center (PARC) supported an in-house artist-in-residence

[8]Florian Roetzer, 1989, "Aesthetics of the Immaterial? Reflections on the Relation Between the Fine Arts and the New Technologies," *Artware,* catalog for the "Kunst und Elektronik" exhibition, Hannover, Germany.

[9]See <http://www.interactiveinstitute.se>. Also see the discussion in the section "Hybrid Networks," below.

program, whose intent, according to PARC director John Seely Brown, was to serve as "one of the ways that PARC seeks to maintain itself as an innovator, to keep its ground fertile and to stay relevant to the needs to Xerox."[10] Some of these efforts have had positive aspects, but few, if any, last very long. Much of the difficulty seems to be how to justify them in the corporate context, especially in difficult economic times. Interval Research is closed and PARC's work is increasingly outside the ITCP domain.[11]

# MULTIFACETED NEW-MEDIA ART AND DESIGN ORGANIZATIONS

This section reviews a few of the best-known examples of new-media art and design organizations, based mostly in Europe (given the lack of such organizations in the United States), in two categories: standalone centers and hybrid networks.[12] Some of these examples might serve as learning experiences for the future establishment of similar U.S.-based organizations.

## Standalone Centers

Among the most prominent standalone centers is the ZKM[13] (a collection of commonly housed institutions involved in new-media art), which encompasses research and development institutes, a media art library, a media art museum, and a modern art museum. The ZKM regularly hosts talks, workshops, and conferences on issues in art history, technology policy, and cultural criticism of technology. A North American version of this model is the Banff Centre in Canada, although the scope of the Banff Centre extends beyond new-media work and its exhibition programs are limited. Other centers include Ars Electronica in Austria, the Center for Culture and Communication (C3) Foundation in Hungary, and the Waag Society in the Netherlands.

Established in 1988, the ZKM (Center for Art and Media) is an influential institution that plays an important role in the German cultural scene (ZKM has hosted the Bambis, for example, the German equivalent of the Emmy awards). Founded on the idea that giving

---

[10]John Seely Brown, 1999, "Introduction," p. xi in *Art and Innovation: The Xerox PARC Artist-in-Residence Program*, Craig Harris, ed., MIT Press, Cambridge, Mass.

[11]See the section "Corporate Experiences with Information Technology and Creative Practices" below in this chapter for further discussion of corporate initiatives.

[12]Also see the discussion of funding organizations within the international context in Chapter 8.

[13]See <http://www.zkm.de/>.

artists access to all of the hardware and software tools and programmers they need to realize their creative vision, ZKM's research institutes maintain cutting-edge hardware and software and a staff of programmers and designers. Because of the depth of resources available to individuals who have a commission or residence at ZKM, artists are encouraged to experiment. The rarity of a (relatively) non-resource-constrained environment being offered by such an organization has resulted in ZKM attracting many established and well-known new-media artists. In addition to supporting art-oriented work, ZKM established two institutes with other foci—one institute with a focus on basic research and another with a focus on socioeconomic research.

Located within the mountains of Alberta, Canada, the Banff Centre hosts work that ranges from the visual arts to theater to music to media. An elaborate resource to facilitate creative thinking and production, the center provides artists with access to studios, technical resources, and trained and skilled production interns. When supporting a project, the center has sometimes been able to supply an artist with a highly trained collaborative team, which includes programmers and designers, to work with the artist until his/her project is completed.[14] Using repeat invitations, Banff has a policy of developing relationships with people over time, and this has become a positive and even essential component of the center's atmosphere. The environment is intense and intimate; in the mountains, people tend to hang out together, working, eating, and socializing. The downside of the residency experience tends to be stretched resources, both human and economic, that make it difficult to create continuous work flow. Technical team members are often working on multiple projects, sometimes making it very difficult to move a project forward and keep it coherent. In addition, the time limits on residencies mean that work must fit within the allotted time rather than developing and resolving at its own rate. How credit is given in teams has also become an issue, as there can be an unfortunate tendency to treat technical collaborators as a kind of service crew. Like ZKM, the Banff Centre has a history of intellectually stimulating seminars and conferences that bring together a broad variety of people from both the non-profit and commercial sectors of new media. At Banff these are often the top people in the field, and the center has become known for the cross-pollination among disciplines that it offers.

Ars Electronica,[15] which is located in Austria, has a rich history of promoting collaborations involving art, technology, and society.

---

[14]The Art and Virtual Environments seminar, for instance, included eight projects experimenting with virtual reality tools. As an example, one virtual reality installation was a murder mystery entitled "Archeology of a Mother Tongue" that involved a collaboration between Michael Mackenzie, a playwright and director working in Montréal, and Toni Dove, an artist based in New York (and a committee member). See Mary Anne Moser and Douglas MacLeod, eds., 1996, *Immersed in Technology: Art and Virtual Environments*, MIT Press, Cambridge, Mass.

[15]For more information, see <http://www.aec.at/>.

Through a series of ventures dating back as far as 1979, it has established itself as a model for other organizations to follow. Ars Electronica consists of four major activities: the Ars Electronica Center, a media center showcasing art and technology projects; Futurelab, the center's research and development arm; Festival, an annual festival focusing on the convergence of art and technology; and Prix, an international competition for "cyberarts."

The Center for Culture and Communication (C3) Foundation[16] in Budapest bills itself as an institution whose main focus is fostering cooperation among the spheres of art, science, and technology, as well as providing a space for innovative experiments and developments related to communication, culture, and open society. Begun in 1996 as a 3-year pilot project, C3 is the result of a cooperative effort of the Soros Foundation Hungary, Silicon Graphics Hungary, and Matáv (a Hungarian telecommunications company). In November 1999, C3— with the cooperation of its founders—became an independent, non-profit institution. One of C3's goals has been to assist with the integration of new technologies into Hungary's social and cultural tradition. Along those lines, C3 has been involved in efforts to provide Internet access to many individuals and non-government organizations, as well as operating a free Internet café and offering free courses regarding how to use the Internet.

The Waag Society[17] was established in 1994 in Amsterdam as a foundation whose research program is focused on the ways in which people express themselves through new and old media. The society has four distinct programs: creative learning, interfacing access, public research, and sensing presence. In addition, the society is also cooperating with partners from industry on several projects. Among these is PILOOT, a communications environment for people with mental disabilities.

Standalone new-media arts and design centers possess unique strengths and face special challenges. Among their strengths, they can offer both stability and freedom to creative people who need work space and technical support. But most such centers are small and resource-constrained. Many centers are based on artist-in-residence programs or colonies with a short-term orientation, and the accumulation of discrete projects with targeted, limited-time funding forms the research agenda. As people and money come and go, organizations may have difficulty sustaining long-term intellectual agendas and pursuing topics iteratively in a way that leads to new insights; therefore, they seldom are viewed as knowledge-building entities like corporations or university departments. They also may have difficultly in acquiring innovative (often expensive, at least for the budgets of many non-profits) technological capabilities. It is possible and desirable, but difficult, for centers to have real research programs with multiyear

> Standalone new-media arts and design centers possess unique strengths and face special challenges.

---

[16]For more information, see <http://www.c3.hu/>.
[17]For more information, see <http://www.waag.org/>.

funding. One approach is to obtain sponsorship from the IT industry; another is to develop durable research partnerships with universities and corporations. Short-term partnerships (on a project basis) sometimes have been successful, as when corporations provide significant resources in a co-development project.

The optimal research strategy is not clear. Short-term programs can be valuable. Artists' residencies are important as a way of democratizing access to technologies and networks of expertise that many artists might not otherwise be able to acquire.[18] But an agenda dominated by discrete projects does not encourage the development of work with depth and richness. The other extreme of very rigid, long-term research agendas is likely no better, because such agendas presume that ITCP work can be clearly defined years in advance—a problematic assumption. At the level of the individual, as noted in Chapter 2, there is a tension between the time needed for playful exploration and conceptualization and the pressure for production. If standalone centers can strike the right balance among these competing pressures, they can fill unique ecological niches where ideas and new art forms are incubated successfully. Moreover, standalone centers assume greater importance in the overall scheme of studio-laboratories in light of the difficulties faced by applied ITCP activities in corporations, which are discussed in the last section of this chapter.[19]

## HYBRID NETWORKS

The term "hybrid networks," which has become popular in some parts of the ITCP community, refers to consortia of diverse organizations. These networks are a means of supporting cooperation and collaboration across disciplines and communities—if they can overcome the communication problems, varying criteria for success, and other difficulties inherent in ITCP collaborations (as outlined in Chapter 2). In Europe, as well as in Canada, hybrid networks have sought to assemble a critical mass of ITCP workers, coupled to industrial sponsors and public policy usually aiming to develop regional innovation

---

[18]Despite the lack of multifaceted new-media arts organizations in the United States, there are artists' residency programs within academia, such as at Arizona State University and the University of New Mexico, and smaller organizations such as Harvestworks in New York City. There are also local programs in many cities that provide access on a more limited basis to computers, skilled teachers, and other resources.

[19]New investments in physical infrastructures have recently been announced, though it is by no means certain how far some of them will be curtailed by economic cycles. In New York City, the Eyebeam Atelier (<www.eyebeam.org>) proposes a venue like the Ars Electronica Center. At Rensselaer Polytechnic Institute, a $150-million-scale Experimental Media and Performing Art Center has been announced. The Presidio development in San Francisco is negotiating a transfer of space to LucasFilm for mixed-use "high tech cultural development." See "Open Letter" to create an Arts Lab—a hybrid art center and research facility—at the San Francisco Presidio, available online at <http://www.naimark.net/writing/artslab.html>.

clusters in creative industries. The decentralized network of technocultural centers supported by the federal government in Canada during the 1990s included the Banff Centre (discussed above), which used this support to initiate its significant investigation of virtual environments in partnership with university researchers and industry sponsors.[20] The European Commission's first round of hybrid research networks was focused on intelligent information interfaces (I3) and linked art centers, corporations, and national laboratories; continued work is being planned to support high-quality digital content and cultural-heritage preservation efforts.[21]

The I3 networks attempt to join art, product design, and academic research and development with corporate product development and market research. In addition, research nsch networks for digital arts began to appear in the 1990s, consistent with a turn away from the stovepipe separation of R&D and commercialization. The concept is captured in the European Commission's Fifth Framework Programme, which emphasizes cross-disciplinary work and features as a theme the "User-Friendly Information Society."[22] The European Cultural Backbone (ECB) mediates between the commercial, government, education, and the cultural sectors, facilitating cross-disciplinary projects by serving as an incubator. The ECB can act as a brokerage structure for mutual support and collaboration to bring together complementary initiatives, drawing upon and connecting local and regional networks of its members in Europe.[23]

In North America, the idea of consortia, networks, and sustained collaboration of this type is attracting greater attention. Examples include the U.S. federal government's Digital Libraries Initiative Phase 2[24] and Internet2 project (both of which include some aspects of ITCP work); the Ford Foundation study in progress, "Art Technology Network" (grant to the Kitchen in New York for a feasibility study, 2001); the activities of the National Alliance for Media Arts and Culture (NAMAC), dedicated to encouraging film, video, audio, and online/multimedia arts and promoting the collaborations of individual media artists;[25] and, at the regional scale, the BRIDGES Consortium (University of Southern California with the Banff Centre). The Canadian government is also supporting regional innovation networks to be admin-

---

[20]Moser and MacLeod, eds., 1996, *Immersed in Technology.*

[21]See <http://www.cordis.lu/ist>.

[22]For information on the Fifth Framework, see <http://www.cordis.lu/fp5/home.html>. For further discussion, see Paul David, Dominique Foray, and W. Edward Steinmueller, 1999, "The Research Network and the New Economics of Science: From Metaphors to Organizational Behaviors," *The Organization of Economic Innovation in Europe*, Alfonso Gambardella and Franco Malerba, eds., Cambridge University Press, Cambridge, U.K., and New York.

[23]See <http://www.e-c-b.net/>.

[24]See Chapter 8 and <http://www.dli2.nsf.gov>.

[25]See <http://www.namac.org>.

istered by CANARIE.[26] Clearly there are opportunities to advance this concept further.

The establishment of additional collaborative networks is one way to increase connections between existing activity in experimental arts and design, cultural and presenting institutions, and the IT industries.[27] But multiple projects are likely to be needed, stressing variously the mobility of people, exchange of know-how, circulation of art and design works, and transfer of technology. Furthermore, all of the institutions involved in such networks need an awareness of the kind of research-application/art-design-technology hybridity exemplified by the Bauhaus.[28]

• • • • • • • • • • • • • • • • • • • • • • • • • • • • • • • • • • • • • • • •

# OTHER VENUES FOR PRACTITIONERS

In addition to multifaceted new-media arts and design organizations, ITCP practitioners can access supportive resources through several mechanisms, principally virtual-space-based strategies and professional conferences.[29] It is interesting to note that traditional arts and design organizations—museums, galleries, art and design fairs, magazines, and so on—are not the prime movers of these efforts. Rather, a patchwork of diverse entities has organized and funded the virtual strategies, and scientific and technical communities organize a number of the conferences.

## VIRTUAL-SPACE-BASED STRATEGIES

A virtual-support infrastructure came into existence in recent years to assist individuals in accessing creative tools and communities, enable communication of various types, and provide shared space for collaboration and file storage and exchange. These organizations have been the most focused of any institutions on leveraging IT to achieve their goals. They tend to be not-for-profit cultural establishments.

---

[26]CANARIE Inc. is Canada's advanced Internet development organization. It is a not-for-profit corporation supported by its members, project partners, and the national government of Canada. CANARIE's mission is to accelerate Canada's advanced Internet development and use by facilitating the widespread adoption of faster, more efficient networks and by enabling the next generation of advanced products, applications, and services to run on them. See <http://www.canarie.ca/about/about.html>.

[27]See <www.thekitchen.org/FordFinal.pdf> for the basis for such a network design proposal.

[28]However, physical proximity still matters, and so hybrid networks that are constrained to a confined geographic area have advantages. See Chapter 7 for further discussion.

[29]And, of course, they have some access to organizations associated with academic institutions; see Chapter 6 for a discussion.

A particular strength of these organizations is that they offer people access to spatially decentralized communities. The result is small, sharply focused groups with low overhead. Some of the better-funded organizations also try to host small offline events for members in an effort to strengthen the online community.

These organizations draw from a common pool of tools to support electronic communication. In general, they use widely available IT, such as directories, databases, online chat programs, discussion boards, and listservs, rather than specialized or state-of-the-art tools. The creation and support of e-mail listservs constitute a common mechanism for organizations that seek to foster community. These tools support, ultimately on a widespread basis, the online equivalent of the gallery, workshop, community center, and coffeehouse, sometimes simultaneously.

Internet-based environments for supporting creative work take many forms, including archives, portals, communities of practice, and virtual galleries. The Internet has made it much easier for grassroots organizations to identify, filter, and collect creative work.[30] Directories (or portals or indexes), usually implemented through Web sites, aggregate pointers to a broad set of resources connected with a domain. Organizations pull together these pointers and then, using a taxonomy they have developed, sort them into categories from which Web pages are then made. For example, the Digital Arts Source (DAS),[31] a "curated" index to new-media art, divides its site into 16 "departments" with titles such as digital art, sound, software, and tools. By entering one of these departments, a user opens a page to a listing of links to resources; links are opened within a new browser window, leaving the DAS site open in the background.

An example of a not-for-profit organization that leverages the use of IT to promote communication between professional and amateur new-media artists is Rhizome.org. Using the Internet as its primary medium, Rhizome carries out its mission of presenting artwork by new-media artists, providing a forum for critics and curators to foster critical dialogue, and archiving new-media art.[32] The Rhizome ArtBase project provides selected Internet artists with dual-purpose server space to store their electronic art as well as the space to make it available for online viewing. Creating a taxonomy using metadata[33] to tag the art based on its type, genre, and the technology used in its production, Rhizome has organized an extensive library of digital art.

A particular

strength of

not-for-profit

cultural

establishments

is that they offer

people access to

spatially

decentralized

communities.

---

[30]These mechanisms can aid in validating and recognizing quality work (see the discussion in Chapter 7).

[31]See <http://www.digitalartsource.com>.

[32]See "what we do" at the Rhizome site, <http://rhizome.org/info>.

[33]Metadata are data about data. In information science, metadata are definitional data that provide information about, or documentation of, other data managed within an application or environment.

The communication and information capabilities of the Internet have allowed Rhizome to break down barriers caused by distance and to build the Rhizome community to more than 13,000 members representing more than 118 countries. To facilitate communication within such a large and diverse group, Rhizome provides a series of moderated and member-supervised digital resources that support many-to-many communication. Internet art news is made available to not-for-profit organizations that seek to provide a news service to their users but do not have the internal resources to do so.[34]

The principal mechanism used by Rhizome to promote asynchronous communication is to push information to individuals by using e-mail listservs—*The Rhizome Digest* and the Rhizome Raw list (equivalent to a conference call that lasts all day every day, this list is unfiltered and enables subscribers to exchange virtually an infinite number of messages almost in real time; messages are stored in a searchable archive). Nettime,[35] a completely virtual service provided by Public NetBase t0, also consists of a series of e-mail lists, most notably Nettime-L. The organization's goal is to provide space for the discussion of new communication technologies and their social implications. Formed in 1995, the listserv has resulted in a conference in 1997 and a book, *Readme: ASCII Culture and the Revenge of Knowledge*, published in 1999.

A number of online communities assist practitioners in ITCP-related fields in locating resources. For example, the Museum Computer Network[36] is a non-profit organization dedicated to fostering the cultural aims of museums through the use of computer technologies. It serves individuals and institutions wishing to improve their means of developing, managing, and conveying museum information through the use of automation. It has hosted conferences with titles such as "Real Life: Virtual Experiences, New Connections for Museum Visitors." The ArtSci project has led to an online resource called the ArtSci Index, intended to create a rich global database of resources and requests for individuals wishing to collaborate, barter, research, or fund collaborative projects involving science and art. A "matching" function assists with the filtering of data relevant to the user's needs. In the architecture arena, ArchNet[37] is designed as an online community for architects, planners, urban designers, landscape architects, architectural historians, scholars, and students, with a special focus on the Islamic world.

---

[34]A syndication model is used in which an Internet art news "module" is added to Web sites for free. This module contains an introduction to a top story and a link to the Rhizome Web site, where the complete story is located. Through the use of HTML to call a JavaScript procedure on the Rhizome Web server, the top story changes daily on the syndicator's Web site without the need for manual interference.

[35]See <http://www.nettime.org/>.

[36]See <http://www.mcn.edu/>.

[37]See <http://archnet.org>.

Some of the organizations providing these services are for-profit. The Thing, an organization similar to Nettime, started operation in 1991 as an electronic bulletin-board system (BBS) for artists and has since evolved into a full-fledged telecommunications company while still maintaining its focus on art and artists. The Thing continues to provide BBS services, functioning as an Internet service provider (ISP) and maintaining and supporting *radar*, a selective calendar of New York art, performance, digital, video, and film events. Individuals and organizations from the art community of New York City contribute submissions to *radar*.

It is possible even for smaller centers and not-for-profit community organizations to access networked technologies that allow for multisite performance. Innovative uses of the Internet and desktop computer technology are producing an explosion of creativity in this area. Funding for smaller grassroots organizations allows the acquisition of both a basic level of technology such as desktop computers, software, and broadband connection possibilities, and the expertise to begin to bring skills to a new population of independent producers. Harvestworks Inc. in New York City is a small, not-for-profit organization with a sound studio, multimedia facilities for production, and classes in a broad area of new technology tools. It runs a Web-based composer's database, a CD magazine, and an artists' residence program that encourages experimental projects and helps to implement them. It has sponsored and supported a number of experiments in multisite performance that have accessed networking tools through collaborations with universities on Internet2 or through smaller theater and performance centers with more modest equipment.

## PROFESSIONAL CONFERENCES

Professional conferences are another means of communicating and displaying the results of ITCP work and may be an ideal way to bring together cross-disciplinary groups of scientists, artists, designers, technical researchers, and others whose paths might not otherwise intersect. There are international symposia that touch upon ITCP, such as ArtSci,[38] which explores how the discoveries of scientific research can combine with the powerful metaphors of art to influence society. Multimedia presentations have featured collaborative projects such as brain waves transmitted via the Internet; the presenters also have discussed issues relevant to ITCP, such as the opportunities and pitfalls of collaborating across disciplines.[39] Another example is the International Computer Music Conference (ICMC), an annual conference

---

[38]See <http://www.asci.org/ArtSci2001/introduction.html>.

[39]The symposium is co-produced by Art & Science Collaborations Inc. and several programs and departments of the City University of New York, with support from the Rockefeller Foundation, AT&T, *Leonardo* journal, and *Nature* magazine.

(since 1974) of the International Computer Music Association.[40] The ICMC includes composers, performers, scientists, and engineers on a roughly equal footing and typically has technical sessions and two or more concerts every day. Papers are selected by peer review, and music is selected by a jury of peers. The 2002 Dokumenta conference included many works that use projective technologies and involved a number of participants from developing countries.

In addition to conferences such as ArtSci, ICMC, and Dokumenta, professional conferences that are relevant to ITCP are also rooted in scientific and technical disciplines. In computer science, the core group of conferences, many of them cross-disciplinary, are organized by professional societies, notably the Association for Computing Machinery (ACM) and the Institute of Electrical and Electronics Engineers, including its Computer Society.[41] Of particular relevance are the ACM's annual conferences for the Special Interest Group on Computer Graphics and Interactive Techniques (SIGGRAPH) and Special Interest Group on Computer-Human Interaction (SIGCHI). See Box 5.2. Specialized conferences related to ITCP are also held under the auspices of technical societies, such as the Creativity and Cognition conference held every third year at Loughborough University, an ACM SIGCHI international conference.[42]

A macro-level approach to communicating and promoting work that crosses disciplinary boundaries is exemplified by the U.S. National Academies' Frontiers of Science and Frontiers of Engineering conferences. The former series, for example, brings together the best young scientists from a variety of fields to discuss cutting-edge research in ways that are accessible to colleagues outside their fields. The idea is to fertilize cross-disciplinary understanding and inspiration by exploring parallels and divergences between different research programs and protocols. One can envision a new venture based on this model called Frontiers in Creativity.

There are also a number of ad hoc conferences sponsored by professional organizations and universities aimed at bringing together scientists, humanists, artists, designers, and engineers to address topics related to IT. Examples include the First and Second Iteration Conferences sponsored by the School of Computer Science and Software Engineering at Monash University in Melbourne, Australia, a juried competition that brought together musicians, artists, and humanists with computer scientists, physical scientists, and engineers to discuss projects recognized for their achievements in using digital technologies creatively.[43] Oriented more toward research, the National Science Foundation, the Andrew W. Mellon Foundation, and

---

[40]See <http://www.computermusic.org>.

[41]There are also specialized professional societies that host relevant conferences, such as the American Association for Artificial Intelligence.

[42]See <http://creative.lboro.ac.uk/ccrs/CC02.htm>.

[43]See <http://www.csse.monash.edu.au/~iterate/>.

---

**BOX 5.2**
**Selected Conferences**

*Special Interest Group on Computer Graphics and Interactive Techniques*

Founded as a special-interest committee in the mid-1960s by Andries van Dam of Brown University and Sam Matsa of IBM, the Special Interest Group on Computer Graphics and Interactive Techniques (SIGGRAPH) was officially formed as a special-interest group of the Association for Computing Machinery in 1969.[1] The SIGGRAPH's premier event, its annual conference and exhibition, brings together artists, scientists, academics, and researchers from around the world working at the intersection of computer science and graphics. The conference has become an important venue, especially for companies producing products that assist the creative process (such as animation software and graphics acceleration cards). One of the more popular experiences is the SIGGRAPH Art Show, a place within the conference where new-media artwork is available for perusal by conference attendees.[2]

*Special Interest Group on Computer-Human Interaction*

The Special Interest Group on Computer-Human Interaction (SIGCHI) is an international group of almost 5,000 specialists working in a variety of disciplines including human-computer interaction, education, usability, interaction design, and computer-supported cooperative work.[3] Of special interest to members of SIGCHI is the intersection of such fields as user interface design, implementation and evaluation, human factors, cognitive science, social science, psychology, anthropology, design, aesthetics, and graphics. A key element of SIGCHI's strategy is the organization and support of a series of cross-disciplinary conferences, many occurring annually. In addition to its main conference, "CHI: Human Factors in Computing Systems" (held annually since 1983), SIGCHI also sponsors conferences in areas such as creativity and cognition (annually since 1999), hypertext and hypermedia (annually since 1987), multimedia (annually since 1993), the design of interactive systems (annually since 1995), virtual reality software and technology (in conjunction with SIGGRAPH and occurring annually since 1997), and other areas where the hardware and software engineering of interactive systems, the structure of communication between human and machine, characterization and contexts of use for interactive systems, methodology of design, and new designs coexist.

---

[1]Information adapted from the SIGGRAPH Web site at <http://www.siggraph.org/>.

[2]While in some respects SIGGRAPH is undeniably successful (e.g., it attracts tens of thousands of attendees to its annual conference), several reviewers and committee members observed that the status of ITCP work and of artists relative to technologists at the conference is ambivalent and perhaps declining.

[3]Information adapted from the SIGCHI Web site at <http://sigchi.org/>. For a photo history of SIGCHI dating back to 1982, see <http://sigchi.org/photohistory/>.

---

Harvard University Art Museums jointly sponsored an invitational workshop in 2001 called "Digital Imagery for Works of Art."[44] The

---

[44]The final report of the workshop is available at <http://www.dli2.nsf.gov/mellon/report.html>. This Web page also contains a link to a comment form seeking feedback, including links to ongoing efforts in the various areas emphasized in the report, as well as pointers to resources (collections, tools, and so on) that may be useful for future collaborative work.

workshop was designed to bring together computer and imaging scientists who have been active in digital imagery research with a particular group of end users, namely research scholars in the visual arts, including art and architecture historians, art curators, conservators, and scholars and practitioners in closely related disciplines. The purpose was to explore how the research and development agenda of computing, information, and imaging scientists might more usefully serve the research needs of research scholars in the visual arts. At the same time, participants looked for opportunities where applications in the art history domain might inform and push IT research in new and useful directions. Other notable examples include the Millennial Open Symposium on the Arts and Interdisciplinary Computing (MOSAIC) 2000 that brought together people interested in the relationship between mathematics and the arts and architecture;[45] the Archaeology of Multimedia conference at Brown University in November 2000; the Re: Play conference in November 1999;[46] the International Symposium on Electronic Art (ISEA);[47] and the Doors of Perception conference.[48]

## PUBLIC DISPLAY VENUES

A variety of physical-place-based (or artifactual) and virtual-space-based techniques and strategies are used for the display of ITCP work for access by public audiences. Exhibitions, performance festivals, and presentations and lecture series seem to be the most common physical-place-based strategies. Museums, galleries, and other entities associated with physical places have featured new-media art for several years. In the United States, the promotion of this art form is guided by many of the established and better-known museums. By constructively adapting their space to include new-media art, both offline inside conventional exhibition spaces and online through the Web, and by commissioning work of this kind, these organizations are experimenting with new frameworks for display and presentation. For instance, since Internet art debuted at the 2000 Biennial, the Whitney Museum in New York City[49] has increasingly supported new media inside its physical location (and on its Web site). Such efforts by traditional art museums are not without critics, who point out the limited allocation of resources or the artistic weaknesses of some shows that may be presented primarily to keep up with trends. Still, some strategies have proven to be both novel and artistically credible.

[45]Held at the University of Washington, Seattle, in August 2000. See <http://www.cs.washington.edu/mosaic2000/>.

[46]See <http://www.eyebeam.org/replay/>.

[47]See <http://www.isea.jp/E/index.html>.

[48]See <http://flow.doorsofperception.com/>.

[49]See <http://whitney.org/>.

The New Museum of Contemporary Art[50] in New York has adapted place-based strategies—both indoors and outdoors—to accommodate ITCP work. The museum features an entire floor where visitors can drop in to read, think, talk, browse, and experience interactive art. Its Zenith Media Lounge is dedicated to the exhibition and exploration of digital art, experimental video, and sound works. This unique space, modest in physical size and budget, draws praise for its adequate technology infrastructure (i.e., computers, projectors, wiring, and so on) and staff support (i.e., system administrators, installers, and others) as well as an artistic focus ensured by staff curators bringing in guest curators to add new-media expertise. A group show, "Open_Source_Art_Hack," which explored hacking practices, open source ethics, and cultural production to provide commentary about life in a networked culture, featured works such as an anti-war game, a packet-sniffing application, and an "ad-busting" project. The museum's Education Department also organizes panel discussions, film and video series, and lectures to introduce challenging new concepts to a broad public.

Physical-place-based displays for ITCP work are also evident at popular technology-oriented museums, such as the Tech Museum of Innovation[51] in San Jose, California. Visitors can make a digital movie using the latest tricks in animation, for example, visit with the roaming robot ZaZa, or engage in high-tech play in the Imagination Playground. A short distance up the coast, Exploratorium: The Museum of Science, Art, and Human Perception,[52] housed in San Francisco's Palace of Fine Arts, is designed to create a culture of learning. A collage of more than 650 blinking, beeping, buzzing exhibits invites visitors to make their own discoveries about the world.

The distinction between physical and virtual display approaches can be blurred in cases in which an organization uses both strategies, and the projects are interconnected—a not uncommon scenario. For example, some museum and gallery exhibitions take advantage of Web technology to enable fluid interchanges between virtual and real spaces. At the exhibition "010101: Art in Technological Times" at the San Francisco Museum of Modern Art (SFMOMA)[53] in 2001, physical installation spaces were supplemented by computers at which gallery visitors could access essays by artists and theorists. Other exhibitions use virtual spaces as part of the artworks themselves, such as in interactive dramas that combine responses by actors in the physical space with online participants who collaborate to produce emergent narratives. See Box 5.3.

---

[50]See <http://www.newmuseum.org/>.

[51]See <http://www.thetech.org/>.

[52]See <http://www.exploratorium.edu>.

[53]See <http://www.sfmoma.org/>.

## BOX 5.3
### Computer and Sensor-based Art Installation

Tiffany Holmes's computer and sensor-based installation, "amaze@getty.org," relies on "liveware"—that is, real-world and real-time viewer collaboration—not just hardware. Created for the exhibition "Devices of Wonder: From the World in a Box to Images on a Screen,"[1] this networked system consisted of tiny spy cameras distributed throughout the gallery space to capture people looking at, and interacting with, other works in the show (such as the multiplying spectacles, book camera obscura, Engelbrecht Theater, and Mondo Nuovo peep show). The live video was ported using wireless technologies into a computer that integrated the imagery into a large montage displayed on a plasma screen encountered as one left the exhibition. The space around the plasma screen acted as the interface for the piece.

In its resting state, with no viewer standing before it, the screen displays a panoramic image of Robert Irwin's garden lying just outside the museum's walls. When a viewer approaches the screen the animation changes. Slowly, the video is fractured by small rectangles. As they multiply, the sweeping image of the garden is replaced by even smaller images grabbed earlier by the hidden cameras. At the end of the piece, layers and layers of imagery fuse and then are broken apart in a simple game of breakout. See Figure 5.3.1.

Conceptually, the piece draws on the metaphor of armchair travel, moving from material terrain into datasphere. The longer the viewer lingers in the gallery, the more she travels virtually, away from the seemingly infinite landscape into a series of highly localized private spaces that ultimately incorporate the watching subject into their confines. Unlike the hyper-illusionism of virtual reality, the experience makes the viewer acutely aware of how visual technology structures perception and how we structure visual technology.[2]

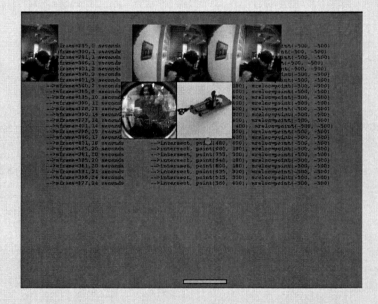

FIGURE 5.3.1 Screenshot from the "Devices of Wonder" real-time animation in which a computer-controlled breakout game removes the specific evidence of the localized surveillance imagery. Contributed by Tiffany Holmes, School of the Art Institute of Chicago.

---

[1]Exhibited at the Getty Museum, Los Angeles, November 11, 2001, through February 3, 2002. See <www.getty.edu/art/exhibitions/devices/choice.html>.

[2]See Barbara Maria Stafford, 2001, "Revealing Technologies/Magical Domains," in Barbara Maria Stafford and Frances Terpak, exhibit catalog, "Devices of Wonder: From the World in a Box to Images on a Screen" in *Cosmos in a Palm*, Getty Museum Publications, Los Angeles.

Many organizations, from museums to professional associations, are eager to explore the possibilities that IT offers for communicating their work to peers and/or public audiences. But until recently, traditional museums and galleries have been slow to provide technological resources or exhibit space to creative talent whose primary medium is the Internet. Their initial online efforts have often focused on supporting activity, such as the digitization of existing collections and the use of the Internet as a medium for providing visitor information, promotion, and virtual gift shops. With that experience and further observation of how artistic expression involving IT is evolving, these organizations are rapidly becoming more open to exploring, supporting, and extending the techniques by which Internet creativity is produced and displayed. Increasingly, it is the strategies being tested, as opposed to whether they are doing any ITCP work at all, that differentiates these organizations from one another.

Examples of such efforts include the Whitney Museum's ArtPort, which provides a portal to digital art and art resources on the Internet and also functions as an online gallery space for commissioned Internet and new-media art.[54] This approach creates a dual strategy of supporting artists directly by commissioning projects, but also promoting the development of the Internet art domain in general by providing artists with cataloged resources. The Dia Center for the Arts began commissioning and exhibiting Web-based work in the mid-1990s.[55] The SFMOMA is expanding its presence on the Internet through the development of its e•space program.[56] With projects dating back to 1996, the e•space initiative supports Internet art via commissions and exhibition. Functioning as a portal to other Internet art or providing a gateway to resources is not one of the project's main goals, although collaboration is. A recent illustration is CrossFade, which concentrates on the Web as a performance space for sound art, an international collaboration involving SFMOMA, the Walker Art Center, the Goethe-Institute, and ZKM.

New-media art is also beginning to make inroads in mainstream galleries, led largely by Internet art and the macro-trend of convergence among media and disciplines. Commercial galleries such as Postmasters[57] and Sandra Gering[58] are now showing and selling new-media art. There is even a commercial gallery dedicated to new media: Bitforms.[59] For an alternate mechanism for matching up artists and collectors, see Box 5.4.

Until recently, traditional museums and galleries have been slow to provide technological resources or exhibit space to creative talent whose primary medium is the Internet.

---

[54]See <http://whitney.org/artport/>.

[55]See <http://www.diacenter.org/rooftop/webproj/index.html>.

[56]See <http://www.sfmoma.org/espace/espace_overview.html>.

[57]The Postmasters Gallery features the work of Perry Hoberman (see Chapter 2) and Natalie Jeremijenko (a committee member). See <http://www.thing.net/~pomaga/>.

[58]The Sandra Gering Gallery presents the work of two individuals featured in this report: Karim Rashid and John Simon (see Chapter 2). Information about the gallery is available online at <http://www.geringgallery.com/>.

[59]See <http://www.bitforms.com>.

> ### BOX 5.4
> ### Mixed Greens
>
> The for-profit organization Mixed Greens does not deal with new-media art specifically, but it has an interesting business model that might be extended to other contexts. Describing itself as an organization that discovers, supports, and promotes artists, this multimedia production company uses its Web site to facilitate communication between artists and collectors. This model allows collectors to get to know artists through a multimedia, personality-based approach to exhibiting art. There are artist documentitos (short videos about the artists), questions that users can answer to match them up with artists, and ways for users to contact artists and learn more about their work or perhaps start a discussion about anything from artistic influences to commissioning a work.[1] One result is The Mix, a Web-based tool for constructing a personal environment for site members to get to know the artists and their work that Mixed Greens supports.
>
> ---
>
> [1]See <http://www.mixedgreens.com>.

ITCP work creates new demands, because the work is often experimental and non-standard, and because the nexus of IT and creativity involves at least a limited breakdown of the separation between audience and performer.

It is important to recognize that ITCP work creates new demands, because the work is often experimental and non-standard, and because the nexus of IT and creativity involves at least a limited breakdown of the separation between audience and performer. Traditional museums and galleries are often poor choices for the display of interactive and/or time-based work (e.g., narratives) because visitors stroll in and out and are not prepared for an experience of duration, nor are these spaces set up for specific start times. The rigidity of theaters with seats bolted to the floor does not work well, either. As a result, gallery spaces and performance and theater spaces need to be rethought to accommodate a less defined and more flexible practice. Theaters with flexible seating, movable partitions, projection systems with scalable, reconfigurable screen systems, and other innovations along these lines are making new work possible. Examples include the new-cinema theater of the Daniel Langlois Foundation for Art, Science, and Technology (DLF) in Montréal, where seats can be fixed or folded under to alter the more conventional cinema setup for special presentations and which has a state-of-the-art digital projection system. The Schaubuehne Theater in Berlin showcases another innovative design and is based on flexibility of scale, audience organization, and media tools for presentation. However, to a large degree, such innovations are not visible to the general public when they visit movie theaters, playhouses or opera houses, symphony halls, or other arts venues, which appear much as they have for the past decades. There would seem to be a rich set of opportunities for new approaches in these traditional venues. However, such new approaches do not replace traditional venues, which

---

**BOX 5.5**
**Burning Man**

In contrast to much of the work done within research laboratories, academic departments, and arts organizations, Burning Man[1] is an intriguing example of art and technology hybridized outside an institutional framework. In many ways, this event, held annually in the desolate environs of the Black Rock Desert in Nevada, is a counterpoint to what happens in the formal context of grants and research budgets, committees, and disciplines. The festival population, which in recent years has numbered almost 30,000 (mostly from the San Francisco Bay Area), includes thousands of engineers from Silicon Valley, from Cisco programmers to old-school hardware hackers. It is also a homing beacon for the West Coast's ITCP community. It is a large-scale undertaking, all coordinated through electronic mail by a self-organizing web of techies, artists, lightning rods, and logistical magicians. The free-spiritedness of the event is belied by an impressive degree of networked organization.

In a matter of weeks, this group of visionaries, tinkerers, and geeks erects not only a small city, with its own roads, sanitation, medical facilities, and electrical grid (all of which are dismantled in a matter of days—these libertarians run their show like a military operation), but also an eye-popping assortment of heavily technological artworks: towering amalgamations of metal and screens and sound feedback devices running artificial-life code, huge and programmable lighting arrays, explosive spectacles, and laser contraptions. Many of these art projects present unprecedented technical challenges, often requiring on-the-spot, midstream innovation (somehow, there are always enough soldering irons, duct tape, fuses, and volunteer laborers to go around—many of the larger installations combine the community effort of barn raising with the specialized expertise of *Mission Impossible*). In a sense, Burning Man embodies the "gift" economy that drives open-source software, with regard to atoms as well as bits. Burning Man also serves as a combination laboratory/audience for a grand techno-artistic experiment.

Perhaps the most salient aspect of Burning Man, for technologists and artists within institutions, is the extent to which social capital can be leveraged at the intersection of technology and creative practices. What Burning Man has in abundance is not financial resources or institutional support or even human capital, in the sense that corporations do (i.e., salaried employees and administrative support). What it has, in abundance, is social capital—the relationships among people that give the event a reason to exist. Burning Man is a phenomenon that emerges from that network, a physical manifestation of the social and creative ties that go back years, sometimes decades. As a way of manifesting the human relationships that bind art and technology, and leveraging those relationships on a large scale, Burning Man is an object lesson for more formal organizations, both in vision and implementation.

---

[1]Detailed information about Burning Man can be found at <http://www.burningman.com>.

---

may well remain valuable places to experience ITCP work. For an example of a distinctive non-traditional venue, see Box 5.5.

Performance art is also going online. Franklin Furnace,[60] a non-profit arts organization that has long supported performance art and other alternative practices in the downtown New York art scene, went digital in the late 1990s and began supporting digital art through commissions and residencies. Also in New York, as part of an exhibit

---

[60]See <http://www.franklinfurnace.org>.

associated with the New Museum, the Surveillance Camera Players performed in front of a Webcam; the audience could view the performance in person at various locations, including the New Museum window on Broadway, or online. The Brooklyn Academy of Music,[61] which gave its first performance in 1861 and has a reputation for changing with the times, has hosted online interactive documentaries that explore and offer insights into the ideological foundation and creation of work presented on stage. For example, "Under_score: Net Art, Sound and Essays from Australia" exhibits the works of nine artists with portals to sonic experimentation. Thematically centered on the body, these projects confront what it means to be sexed up, desired, gendered, and identified under electronic conditions. Another documentary chronicles the development of the work Love Songs by choreographer David Roussève and his company REALITY, and presented as part of BAM's 1999 Next Wave Festival. Users have access to rehearsal footage, interviews with Roussève, company information, and discussions about the work.[62]

Some experiments with art presentation provide venues for Internet-based research that push the envelope in terms of how technology can be used to augment social environments. The Museum of Modern Art (MOMA)[63] in New York City has supported a number of new-media artists through commissions and has used the Internet to engage artists to overcome obstacles linked to space and time. For instance, its Conversations with Contemporary Artists program, originally presented offline, recently was replicated online. Projects constrained by time such as Time Capsule have been migrated online as well.

Similarly, the Walker Art Center is focusing not on how virtual space enables the display of content, but rather on how it facilitates information sharing and collaboration in the creation of art—a kind of Internet-based creative research and development.[64] Built in 1971, the center concentrates on supporting the development and exhibition of modern art. By supporting film/video, performing, and visual arts, the center takes a global, cross-disciplinary, and diverse approach to the creation, presentation, interpretation, collection, and preservation of art.[65] The new-media initiatives of the center seek to achieve two goals: Its Gallery 9 project promotes project-driven exploration through digital-based media, whereas the SmArt project concentrates on information architecture and the role that it might play in facilitating access to the collections and activities of the center.

---

[61]See <http://www.bam.org>.

[62]One reviewer, however, suggested that BAM is pulling back from such digital initiatives.

[63]See <http://www.moma.org/>.

[64]See <http://www.walkerart.org/jsindex.html>.

[65]See <http://www.walkerart.org/generalinfo/>.

It is important to note that efforts to collect and display ITCP work run into the formidable challenges posed by digital technology. (See Chapter 7.) For instance, the likely inability to distinguish between an original and a copy will have a profound effect on museums, one rivaling the effect that photographic reproduction had on art.[66] This will cause a paradigm shift in how a museum views its holdings (as pimarily unique original objects) and how it certifies their authenticity. Conservationists will also have to shift from the paradigm of repairing and saving a physical object to that of maintaining a set of disembodied artistic content over time. Indeed, it has been said that new-media art questions the most fundamental assumptions of museums: What is a work? How do you collect? What is preservation? What is ownership?[67]

• • • • • • • • • • • • • • • • • • • • • • • • • • • • • • • • • •

## CORPORATE EXPERIENCES WITH INFORMATION TECHNOLOGY AND CREATIVE PRACTICES

A number of industries depend heavily on, and derive substantial profits from, ITCP activities. This may be most obvious in entertainment, or "content-based" corporations. For example, filmmaking has driven and adopted advances in both technical and creative achievement, and computer games would not exist at all without ITCP (the collaborative nature of these industries is discussed in Chapter 2). Yet technology companies seem to be less successful in importing and applying art and design knowledge to their activities. Some companies have experimented with specialized arts centers, artist-in-residence programs, or short-term arts projects, with mixed results.

Two less than successful initiatives were mentioned earlier in this chapter: Interval Research and Xerox PARC's Artist-in-Residence program. Lucent's one-of-a-kind collaboration with the Brooklyn Academy of Music, which produced the acclaimed Listening Post (described in Chapter 2), is now defunct, a victim of the company's financial woes. Tough economic times in 2001 and 2002 have led to the downsizing of the Nippon Telegraph and Telephone InterCommuni-

---

[66]Walter Benjamin, 1969, "The Work of Art in the Age of Mechanical Reproduction," *Illuminations*, Hannah Arendt, ed., Schochen Books, New York; Howard Besser, 1987, "Digital Images for Museums," *Museum Studies Journal* 3(1): 74-78; Howard Besser, 1997, "The Changing Role of Photographic Collections with the Advent of Digitization," *The Wired Museum*, Katherine Jones-Garmil, ed., American Association of Museums, Washington, D.C., pp. 115-127.

[67]As characterized by Jeremy Strike, director, Museum of Contemporary Art, Los Angeles, in *Museums and New Media Art* by Susan Morris, a research report commissioned by the Rockefeller Foundation (mimeo), October 2001.

cation Center (ICC)[68] and folding of the Canon ArtLab,[69] both in Tokyo.

But some IT companies have experienced ventures into art and design that generate qualitative and tangible benefits. The Philips Vision of the Future is one example of a large-scale, successful project that integrated technology development with design practices and human-centered disciplines such as anthropology.[70] (See Box 5.6.) Another example is IBM Corporation's Computer Music Center (CMC), which from 1993 to 2001 focused on the intersection of computer science and music as a research testbed, beginning with an effort to develop the underlying technology for KidRiffs, a consumer software product that was eventually marketed and sold. Over time, the center grew to focus on interactive music composition tools, seeking to understand the special demands that large-scale creative endeavors, such as film scoring, place on software. Research on visual programming languages, interactive graphics, real-time systems, and audio identification provided a diverse backdrop for the core music composition work. Although the CMC was ultimately closed, it contributed to product development, public relations, and the company's portfolio of intellectual property.[71] The QSketcher system, for example, developed in close collaboration with faculty and composers affiliated with the Berklee College of Music, pioneered several HCI mechanisms.[72] The

---

[68]Established in 1997, ICC sought to "encourage the dialogue between technology and the arts using a core theme of 'communication'" and "to become a network that links artists and scientists worldwide, as well as a center for information exchange." Both a museum open to the public and a research lab, ICC was a place for the conceptualization, introduction, and testing of advanced media art communication technologies. But ICC was more than just a setting where artists and technologists congregated to create media art to be displayed in a museum or gallery. Offering workshops, performances, symposiums, and publishing, the center went beyond exploring the possibilities of new forms of communication to attempt to educate and expose the public to the capabilities of the art and technology intersection. See <http://www.ntticc.or.jp>.

[69]ArtLab was a corporate lab focused on the integration of the arts and sciences, primarily by encouraging new artistic practices using digital imaging technologies. The "factory" (laboratory) employed computer engineers using Canon digital products in interaction with artists in residence to produce new digital art works. The studio portion of the program has presented exhibitions of the works developed in-house. In 1995, seeking to introduce multimedia works to the general public, the ArtLab began its Prospect Exhibitions program, which also provides access to the work of multimedia artists and creators outside the ArtLab. Workshops and lectures on new communications technologies and practices were also organized on an ad hoc basis and are both national and international in scope. See <http://www.canon.co.jp/cast/artlab/index.html>.

[70]See <http://www.design.philips.com/vof/toc1/home.htm>.

[71]Sometime prior to the closure of the center, IBM determined that a venture into the music software tools market was unlikely but continued to support the center as a fundamental research effort.

[72]See Steven Abrams, Joe Smith, Ralph Bellofatto, Robert Fuhrer, Daniel Oppenheim, James Wright, Richard Boulanger, Neil Leonard, David Mash, and Michael Rendish, 2002, "QSketcher: An Environment for Composing Music for Film," pp. 157-164 in *Proceedings of the Fourth Conference on Creativity and Cognition*, Loughborough University, U.K., ACM Press, New York.

center hosted well-known musicians, served as a recruiting tool, and contributed to education and outreach programs. The venture also produced a stream of inventions, including 16 patents filed over the center's last 4 years.[73]

Such interactions can also be valuable from the artists' perspective. For example, AT&T's Bell Labs, later partly spun off with Lucent, never had a formal artist-in-residence program, but Bell Labs had some informal arrangements with musicians and filmmakers who were invited to try out software programs, collaborate on specific projects, and provide consulting or video documentation services. No money was provided for music and art projects, although film production expenses were sometimes covered by the parent company (AT&T or Lucent) or outside grants. A committee member who participated in these activities, Lillian Schwartz, recalls that the artists were thrilled to have the chance to work with great minds and unique technologies, and the scientists were inspired by the interactions to produce new and enhanced hardware and software such as a hands-free telephone. In addition, such interactions afford artists and designers with the possibility to have wide-scale impact, by affecting the design of products used on a mass-market scale.

Whether the types of benefits detailed here are sufficient to convince technology companies to initiate and retain artistic/design programs is open to debate. Within a for-profit corporation, there is an understanding that initiatives have to justify their own resource commitments in an expected financial return in terms of some measure such as productivity improvement, cost savings, or increased revenue. Unfortunately, ITCP work, like corporate R&D more generally, does not lend itself very easily to calculations of return on investment or net present value.[74]

Nonetheless, these programs can be evaluated, and their impact ascertained, in a qualitative fashion, much as the impact of an employee can often be ascertained in a qualitative fashion. The questions include the following: Did this project enhance the capabilities of the organization? Did it prompt useful discussion and action? Did it matter to anyone other than the artist or designer? If so, how? If the project is undertaken as a strategic exercise, rather than a public relations exercise, then it has to demonstrate some relevance, even if it does not directly result in a commercial product (as most R&D projects do not).

---

[73]The discussion of the IBM Computer Music Center is based on the personal account of a committee member.

[74]As an aside, it is worth noting that basic research—in the sciences or otherwise—is also very difficult to describe in terms of quantitative benefits, inasmuch as the benefits may become apparent after a considerable lag or may be realized outside the performing unit. Although some companies (e.g., IBM, Lucent Technologies) use counts of outputs such as patents, and most companies measure how much they spend on research-related activities (usually development), such indicators are still primitive and incomplete when it comes to assessing the value of the work.

## BOX 5.6
### A Vision of the Future Pays Off in New Knowledge and Products

The goal of the 1995 Vision of the Future, a project in the design division of a large electronics corporation, was to incorporate quality-of-life considerations into the design of new technological products. Participants created conceptual designs for electronic consumer products, based on scenarios sketching out technological and cultural changes likely to happen in the near future. Although IT products are often driven largely by technical considerations, Philips hoped to demonstrate that its products are intended to improve quality of life;[1] enhance creativity within the Philips organization; use an integrated technical and sociocultural perspective as an opportunity to create new, unique types of products; and demonstrate that a focus on quality of life could have tangible benefits in the form of new product design.

The project created concrete instantiations—models, simulations, and movies—of scenarios developed by cross-disciplinary teams whose members ranged from anthropologists to engineers to graphics designers, based on trends identified by futurists and engineers. These instantiations were used to get user feedback through an exhibition, book, video, and Web site, and to influence the design of future products.

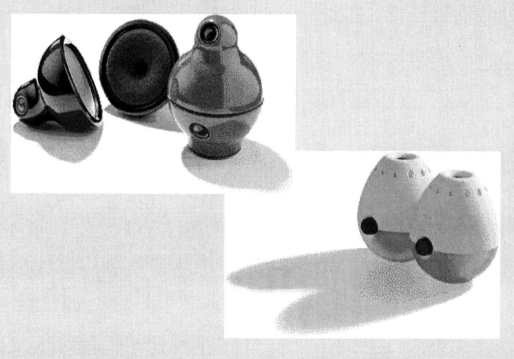

FIGURE 5.6.1 "Emotion containers" come in various personalized, aesthetically pleasing forms that contain smells, sounds, and images and can be given as sensuous and emotionally meaningful presents. Images contributed by Philips Design.

For example, in the future-home scenario, technology is largely invisible, made transparent by being integrated directly into the home and its contents. Screens, for example, disappear, replaced by "living wallpaper"—wall-size, flat displays that could present anything from static or moving decoration to information to entertainment. Electronic objects no longer look "techie," instead featuring sculptural shapes and aesthetically chosen forms that can be personalized to the user. There is an emphasis on technologies that support social connections; an example is Remote Eyes—small, wireless video cameras that can be placed anywhere, allowing people to share images from their daily lives and supporting a variety of playful uses. Environmental values are reflected in technology such as the Intelligent Garbage Can, which automatically sorts trash for recycling, compacts it, and removes its smell.

The Vision of the Future was clearly a design success—it has achieved wide recognition, including the 1996 Design Distinction for Concept Design by *I.D. Magazine* (U.S.). But what about its impact on technology products? According to the Philips Web site, in the last 6 years "60% of all concepts envisioned in The Vision of the Future became actual solutions." Wearable electronics, for example, which were envisioned in the project, are now being produced and marketed in the Industrial Clothing Division line, a partnership between Philips and Nike. The value of the methodology for the company itself is reflected in the development of a series of follow-on projects including the Home of the Near Future (1999), Connected Pl@net (2000), and Smart Connections (2001), each of which analyzes different technologies and application domains using the Strategic Futures methodology pioneered by the Vision of the Future. The value to Philips seems clear.

At the same time, the question of the broader value of human-centered input from anthropologists, futurists, and designers for IT product development remains open. The project's claim of supporting the environment seems to be belied by future scenarios describing the proliferation of inessential consumer goods and personalized (and hence nonreusable) products. At the same time, critics like Genevieve Bell and Joseph Kaye argue that the human values embodied in the Vision of the Future products are not substantially different from those of previous products—that the Vision of the Future manifesto's claim that human values matter more than efficiency is violated by many of the resulting efficiency-focused products, from the Shiva personal organizer to interactive jewelry that connects wearers to agendas and reminds them of appointments.[2] In fact, in describing products specifically intended to promote quality of life, the first attribute of this quality is said to be "when products or services reduce the time or amount of tasks needed to be performed."[3] Issues such as these underscore the need to negotiate the tension between the constraints on Philips as a company in the marketplace—the need to sell more products to consumers that are recognizable to them as consumable objects—and the humanist vision of the designers, ethnographers, and other participants in the project. In the end, claims to be able to shape culture in positive ways cannot be truly evaluated until the products have entered people's homes and daily lives. Clearly, the Vision of the Future is good for the Philips Corporation; it remains to be seen if it will be good for society.

---

[1]For this project, quality of life was understood with respect to two broad issues. The first issue was sustainability, in terms of both environmental sustainability (for example, the materials out of which the products were made) and social sustainability, or the support of an individual's social and emotional needs within society. The second issue was general quality-of-life issues, with a focus on values and aesthetics over efficiency, a traditional focus of electronic appliance design (see, for example, Figure 5.6.1).

[2]See Genevieve Bell and Joseph "Jofish" Kaye, 2002, "Designing Technology for Domestic Spaces: A Kitchen Manifesto," *Gastronomica* 2(2): 46-62.

[3]See <http://www.design.philips.com/vof/vofsite6/index.htm>.

Even if it stirs controversy, or galvanizes opposition (some would say this is a sign of success), the project has to have an impact on the corporate discourse. If it is acclaimed outside the organization and ignored within, that, too, is a point of departure for the institution.

Thus, the argument for introducing outsiders, such as artists, writers, independent designers, directors, curators, activists, anthropologists, and the like, into an IT corporation must be made in the absence of specific supporting empirical evidence. When would such outside perspectives be most valuable? The motivations of outside artists and designers, in the right context, can drive technology forward in unexpected ways. Artists reconfigure unlikely materials for unexpected purposes.[75] Designers understand that commercial products alone do not determine their value as professionals; they often sustain elaborate extracurricular work to demonstrate their creativity. The opportunity cost of risk is tiny in the realm of cultural production, compared with the corporate sphere. And the rewards for risky new ideas are high. These drivers create a strong set of incentives to break new ground by confronting emerging issues,[76] not the least of which is the form, function, and significance of new technologies, and the practices and assumptions that drive technology-oriented organizations. Essentially, the corporation that engages outsiders leverages incentives outside the economic sphere to generate intellectual capital that might not otherwise take root inside the organization.[77]

The primary value of this engagement, from the corporate perspective, is not the production of a better mousetrap—corporate R&D does a good job of optimizing mousetraps—but the discovery of new relevance for the mousetrap.[78] The engagement of outsiders can generate value from the design of more stylish, usable, and aesthetically

> Outside artists
> and designers, in
> the right context,
> can drive
> technology
> forward in
> unexpected ways.

---

[75]See, for example, Vicki Goldberg, 2001, "Industry and Art: A Long Embrace," *New York Times*, April 22, available online at <http://query.nytimes.com/search/abstract?res=F20A17FB3D540C718EDDAD0894D9404482>.

[76]The incentives are intellectual and social—curiosity and glory—rather than commercial. The rewards in this context are overwhelmingly about status, rather than money—an artist's career is, for better or worse, measured by and dependent on awards, inclusion in public and private collections and in museum and gallery exhibits, and critical attention. Architects and designers sometimes earn as much (or more) acclaim for their conceptual designs as for commercial products and real-world buildings.

[77]As economist Richard Caves has argued, "Creative goods and services, the processes of their production, and the preferences or tastes of creative artists differ in substantial and systematic (if not universal) ways from their counterparts in the rest of the economy where creativity plays a lesser (if seldom negligible) role." See Richard E. Caves, 2000, *Creative Industries: Contracts Between Art and Commerce*, Harvard University Press, Cambridge, Mass., p. 2.

[78]New relevance may emerge over time. For example, British Petroleum has begun to investigate applications in medicine for the visualization and virtual reality tools it developed for oil and gas exploration. And a reviewer noted that Schlumberger was the first oil company to get into the medical business as it applied its knowledge of capillary action on another scale.

pleasing products, if indeed it is true that technology has advanced sufficiently that the major differences among products discerned by consumers seldom are based on technological capability per se.[79] The economic payoff of creative practice tends to be at the design level—for example, design has been so important at such companies as Apple Computer and Nike as to be a core competency.[80] Moreover, current trends in ITCP suggest the potential for creative practice to shape IT that is more usable and better integrated with its users than has often been the case—whether among companies that focus on producing IT or on its intense use.

The crux of the argument to bring in outsiders is that the organization's regular staff is unable, on its own, to avail itself of the opportunities articulated above. Outsiders can help to challenge the status quo,[81] to foster cognitive diversity.[82] Resistance to change or inertia often derives from an organization's culture and legacy products—there are proven ways to achieve success, and so it is understandable that the organization's employees can be ambivalent about pursuing unproven avenues. And, indeed, the purpose of many organizations is to perform particular tasks repeatedly, gaining competitive advantage and providing value to the consumer through cost economies.[83]

Some corporations have initiated artist-in-residence programs that have tended to operate at the periphery of the organization. Such programs may produce greater benefits if they are managed as serious endeavors aimed at core functions that can yield important results across the enterprise.[84] (Companies that have terminated programs that operated on the periphery may have had unrealistic expectations for them.)

---

[79]See Vikas Bajaj, 2002, "Makers of Electronics Begin to Emphasize Style," *Dallas Morning News*, May 2.

[80]Of course, the contribution of designers reflects the work of artists, much as the contribution of engineers reflects the work of scientists.

[81]One criterion for the selection of artists or designers in this role is that they will be willing to engage corporate staffs actively. They must not be aloof.

[82]See Dorothy Leonard-Barton, 1995, *Wellsprings of Knowledge: Building and Sustaining the Sources of Innovation*, Harvard Business School Press, Boston, Mass.

[83]Whether an organization bakes bread, manufactures automobiles, teaches students, or provides hotel accommodations, what some might construe to be organizational (or structural) inertia may be characterized as core competence by others. More on this point can be found in the literature on organizational ecology; for example, see Michael T. Hannan and John Freeman, 1989, *Organizational Ecology*, Harvard University Press, Cambridge, Mass; and Glenn R. Carroll and Michael T. Hannan, 1992, *Dynamics of Organizational Populations*, Oxford University Press, New York.

[84]One comparison might be to the efforts of companies such as Apple Computer to designate roaming specialists as evangelists, such as its use at one time of Bruce Tognazzini as its Evangelist to Human Interface to promote attention to usability across the company.

There could be irresistible pressure to institutionalize outsiders into an organization—especially if the outsiders are successful. Formal structures, titles, budgets, and other processes may evolve. Such developments may be essential to some degree, but this institutionalization would have to be monitored carefully. Formalization may undercut the effectiveness of the outsiders if priorities shift toward the protection of budgets and outsiders are enculturated to become insiders. To ensure fresh perspectives, it may be that the tenure of outsiders should be limited to a relatively short period of time.

# 6 | Schools, Colleges, and Universities

School of art and design, colleges, and universities are fertile ground for fostering work in information technology and creative practices (ITCP). As sources of knowledge, education, and training, they facilitate the acquisition of new and different skills and insights. In addition to providing a place for exploring new types of activities and new mixes of skills and knowledge, they bridge old and new knowledge and techniques and ways of thinking and doing, providing a place to see (and study) how the new builds on—and differs from—the old. As the institutional home for students, they provide a ready source of talented and motivated labor to support work in ITCP—a practical reality that encourages some ITCP practitioners to be involved with such institutions.[1]

Academic environments are designed to enable broad, deep, and long-term creative explorations, but how they embrace change, whether within a discipline or in a multi- or transdisciplinary activity, varies enormously. Some generate new programs and embrace new areas readily, while others find programmatic and structural change more difficult, given the challenges and/or inertia in leadership, institutional culture, and the allocation of resources. Approaches to cross-disciplinary opportunities—which today come from many directions—vary, depending on the institution's relative emphasis on research or teaching, the seniority and size of its faculty, and faculty members' willingness and ability to collaborate across disciplinary boundaries.

ITCP activities have begun to proliferate in academia—more visibly and vigorously in departments of art and design than in computer

---

[1]This observation was made by briefers to the committee and is consistent with the experience of some committee members as well.

[2]Computational science involves the application of computer science to study scientific problems, which often involves high-performance computing.

science departments. This imbalance of interest is not necessarily bad; it is analogous to the linkage of computational science[2] programs to other sciences (e.g., computational physics or chemistry) rather than to computer science. In particular, ITCP can be seen in the emergence of various new-media (or digital media or digital arts) activities. These activities relate to ITCP in name, but they appear to vary greatly in their intellectual rigor and vigor and in their impact on preexisting programs in the arts and design.

This chapter begins with a presentation of the specific organizational mechanisms that directly support and promote work in ITCP. Efforts within the mainstream schools and departments—of computer science, art, and design—to advance ITCP work are then discussed. The chapter concludes with cross-cutting observations and implications.[3]

## ORGANIZATIONAL MODELS FOR SUPPORTING WORK

There are three broad categories of academic ITCP organizations: specialized centers, workshops, and service units. Specialized centers are of the greatest interest in the present context because they tend to produce work that balances and integrates disciplines at the deepest levels and for the longest periods of time. However, workshops and service units can also make valuable contributions by working or fostering work across disciplines, as detailed below.

### SPECIALIZED CENTERS

The specialized center is the most visible model for academic work in ITCP. The (relatively) standalone type operates largely autonomously from mainstream academic departments, and the derivative type obtains significant funding from one or more mainstream academic departments within one or more universities. And, of course, there are various gradations of specialized centers between standalone

---

[3]Although this chapter focuses on higher education, the committee does not intend to suggest that ITCP work occurs only in higher education and/or that preparation for ITCP work in colleges and universities cannot begin earlier at the K-12 level. While it neither possessed the credentials to speak authoritatively on K-12 education nor heard testimony from K-12 researchers or educators, the committee does wish to record its sense that ITCP work could have a considerable, positive influence on the curricula of primary and secondary schools. Moreover, rich offerings in the arts and design areas, in addition to mathematics and science, in K-12 education can serve as an important baseline for ITCP thinking and work in the undergraduate years.

and derivative. Representing points along a spectrum, the specialized centers described here are often positioned according to their links to their host academic communities. Specialized centers can serve as a cooperating unit for joint appointments of faculty and staff in mainstream departments, which can encourage intellectual cross-fertilization.

In contrast with more decentralized attempts to evolve existing academic units or to incorporate ITCP elements within traditional units, the standalone center aims to focus on ITCP work from the outset. Experts skilled in one or more areas are convened in a single organizational unit to conduct work in ITCP. Many of the concepts and principles of studio-laboratories apply to academic standalone centers,[4] although the context is different.

The Media Lab at MIT may be the best-known example of a standalone ITCP research center, at least within the United States. It opened its doors in 1985, growing steadily from an interest in computation at the Massachusetts Institute of Technology School of Architecture and Planning and expanding its scope to include multiple arts and rather different fields, such as materials science, which it incorporates through its own hiring as well as joint appointments. It has become both one of the largest centers focused on ITCP[5] and the subject of discussions about whether it should be considered a model, given the mixed success of larger, enduring organizations in the ITCP space in general[6] and some of the specific problems it has encountered in its own growth.[7] Areas of interest include the computational properties of physical systems of all types; the overlay of digital information on the physical world (e.g., tangible media, discussed in Chapter 4); and global outreach, especially to developing nations.[8] The challenge for the Media Lab is to sustain a high level of energy and success as it continues to mature and grow, and to avoid losing its edge by allowing past successes to dominate plans for the future. At present, the model has attracted interest in other countries (e.g., India), where sister facilities are contemplated.

An example of a derivative center is the newly established Center for Research in Interactive, Telematic, and Immersive Culture (CRITIC) at the University of California at Irvine, a cross-disciplinary graduate program in the arts, computation, and engineering that is jointly supported by the School of the Arts, the School of Engineering, and the

> ITCP activities have begun to proliferate more visibly and vigorously in departments of art and design than in computer science departments.

---

[4]See Chapter 5 for a discussion of studio-laboratories.

[5]As of 2003, the Media Lab has about 30 faculty members and senior research staff, 170 graduate students (somewhat evenly divided in master's and doctoral programs), and about 150 undergraduate students.

[6]See Chapter 5.

[7]For further commentary on the Media Lab, see David H. Freedman, 2000, "The Media Lab at a Crossroads," *Technology Review* 103(5): 70-79, available online at <http://www.technologyreview.com/articles/freedman0900.asp>.

[8]For additional information about the Media Lab, a unit of the School of Architecture and Planning at MIT, see <http://www.media.mit.edu>.

(soon to be) School of Information and Computer Science, with additional support from the School of Humanities. Course materials are printed with the disclaimer "may cause permanent damage to axiomatic assumptions."[9]

The CRITIC will be complemented by a research program in a cross-disciplinary, multientity project, the California Institute for Telecommunications and Information Technology (Cal-(IT)$^2$).[10] Although centered on science and technology, Cal-(IT)$^2$ has hooks into the arts and social sciences, and accordingly, it has a range of institutional links. A New Media Arts "strategic application" component[11] focuses on computer games and visualization environments "to provide creative research challenges likely to have impact on distance learning, collaborative work environments, and understanding large complex data sets."[12] It, in turn, involves the Center for Research in Computing and the Arts (CRCA) at the University of California at San Diego, an "organized Research Unit . . . whose mission is to foster advanced research and production at the crossroads between digital technology and new art forms."[13] CRCA activities include interactive networked multimedia, virtual reality, computer-spatialized audio, and live performance techniques for computer music and graphics; it also explores artists' software systems. Cal-(IT)$^2$ is an ambitious experiment in holism and maximal crossdisciplinary effort. Like other large and multifaceted academic centers that build on preexisting components, it is hard, especially at the early stages, to understand how much true substantive integration is actually being achieved.

Whereas Cal-(IT)$^2$ spans two universities in one state, other derivative centers have broader geographical reach (although this remains uncommon). An example is the Graphics and Visualization Center that involved Brown University, the California Institute of Technology, Cornell University, the University of North Carolina at Chapel Hill, and the University of Utah. One of the earliest National Science Foundation-supported science and technology centers for IT, the now-defunct center explored options for future interactive graphical environments, from algorithms to user interfaces, and it was itself a laboratory for tools to support remote collaboration. Additional specialized ITCP centers are described briefly in Box 6.1 to further illustrate the nature of these centers.[14]

> The standalone center aims to focus on ITCP work from the outset.

---

[9]As described to the committee by the center's founder, Simon Penny.

[10]See <http://www.calit2.net/research/index.html>.

[11]Within Cal-(IT)$^2$, this component has a special name, "Arts Layer."

[12]See <http://www.calit2.net/art/index.html>.

[13]See <http://www.crca.ucsd.edu/>.

[14]One important dimension that can be used to characterize specialized centers is whether one world view is dominant (and if so, which one). For example, the Electronic Visualization Laboratory is closer to the computer science field than to the arts or design world, while the Institute for the Study of the Arts is closer to the arts and design world than to the field of computer science.

## WORKSHOPS

Flexible, informal means of promoting truly creative practices are needed, because any institution designed to support some particular conception of creative practice (such as a specialized center) will discourage movement beyond that paradigm, increasingly so as the institution becomes more established. In many universities, workshops, seminars, and other ad hoc convenings provide a quasi-informal meeting ground for testing multi- and transdisciplinary work focused on a knotty problem or a complex area. Such flexible, low-cost mechanisms for bringing disparate people together are inherently bottom-up, intimate, and conversational; typically, drafts of papers are read and discussed, and students, faculty, and visitors mingle across disciplinary divides. At the University of California at Los Angeles, for example, an cross-disciplinary initiative with the acronym SINAPSE (Social Interfaces and Networks in Advanced Programmable Simulations and Environments) provides a variety of forums to encourage discussion and debate of IT issues. This approach has proved effective in encouraging conversations among humanists, scientists, and artists.

There is also much potential for ITCP in a different type of workshop, harking back to the medieval, hands-on space for technological experimentation and testing of thoughts. Part alchemical laboratory, part well-stocked high-tech site, a media lab is the less formal, less expensive correlate to the full-fledged research center. It offers faculty and students a dedicated place for experimentation and an opportunity to produce the kinds of results that can happen only with new media. Media labs may be constituted temporarily, perhaps based on a theme, for periods of 3 to 6 weeks—long enough to permit doing new intellectual and technical work and to support identifying viable problems and projects, but still without imposing the burden—and risk—of making permanent institutional arrangements. Participants can sign up and work anonymously and/or collaboratively, but without fanfare—much as members of the academic community, in the days before the personal computer, flocked to shared computing centers. McMaster University's Multimedia program offers such a facility for faculty and, with the aid of a large grant, is in the process of developing open-source software with collaborative input from a variety of faculty at McMaster and other Canadian institutions. The University of California at Santa Barbara, in conjunction with its Transcriptions project, has also established a high-tech meeting place to encourage collaborative and creative work on curriculum and research projects that incorporate IT.[15]

Workshops and other ad hoc convenings provide a quasi-informal meeting ground for testing multi- and transdisciplinary work focused on a knotty problem or a complex area.

---

[15]From the project Web site (<http://transcriptions.english.ucsb.edu/about/index.asp#concept>): "The goal of [Transcriptions] is to demonstrate a paradigm—at once theoretical, instructional, and technical—for integrating new information media and technology within the core work of a traditional humanities discipline. Transcriptions seeks to 'transcribe' between past and present understandings of what it means to be a literate, educated, and humane person."

## BOX 6.1
### Selected Specialized Centers

• The *Electronic Visualization Laboratory* (EVL) at the University of Illinois at Chicago (UIC), created in 1973, is a graduate research laboratory specializing in virtual reality and real-time interactive computer graphics. EVL's current research areas include scientific visualization, new methodologies for informal science and engineering education, paradigms for information display, distributed computing, sonification, human-computer interfaces, every-citizen interfaces, and abstract mathematical visualization. A joint effort of UIC's College of Engineering and School of Art and Design, EVL offers graduate degrees (M.F.A., M.S., Ph.D.) in electronic visualization. For more information, see <http://www.evl.uic.edu/>.

• The *Institute for Studies in the Arts* (ISA) at Arizona State University (ASU; Herberger College of Fine Arts) supports individual inquiry and collaboration among artists, scholars, and technologists. The ISA sponsors both faculty and student research as well as residencies for visiting artists and scholars. In the fall of 2002, the ISA began offering two new training programs, Interdisciplinary Digital Media and Performance (IDMP) and Signal Processing and Programming for the Arts (SPP). The IDMP program features "survey, lecture and laboratory exposure to the uses of technology as an essential part of cross-disciplinary art and the principles of collaborative art making through collaboration between creative artists and technologists." The SSP program focuses on tools for digital and media arts, incorporating basic signal-processing technical concepts taught for non-science majors. Overall, the ISA program offers as research environments digital imaging, audio, and intelligent stage laboratories, as well as a technology development studio, a collaboration with ASU's College of Electrical Engineering. For more information, see <http://isa.asu.edu/flash_home.html>.

• The *Institute for Creative Technologies* at the University of Southern California (USC) draws on USC's School of Cinema-TV, the School of Engineering and its Information Sciences Institute (ISI) and Integrated Media Systems Center (IMSC), and the Annenberg School of Communication. The institute has a specific mission—meeting the Army's interests in better tools for immersive training and simulation, which may prove to constrain its approaches as well as its projects in terms of ITCP work. It is an experiment in both cross-disciplinary exploration and academic engagement in meeting military needs.[1] Its assumptions are that "[t]he entertainment industry brings expertise in story, character, visual effects, gaming and production. . . . In addition the computer science community brings innovation in networking, artificial intelligence, and virtual reality technology." For more information, see <http://www.ict.usc.edu/>.

• The *Center for Advanced Technology* (CAT) at New York University draws from a larger state program of support for centers of advanced technology. Its mission involves support for New York's new-media industry—the development and licensing of technologies relevant to that industry is a goal. Graphics (and related computer science) is central to the ITCP work undertaken. A committee member associated with CAT joined with another committee member to initiate a local lecture series inspired by this project, MeAoW! (Media Art or Whatever), to facilitate artists' engagement with technologies and technologists. For more information, see <http://www.nystar.state.ny.us/nyu.htm> and <http://cat.nyu.edu>.

---

[1]Another program that addresses military needs is the MOVES Institute, based at the Naval Postgraduate School, which is a cross-disciplinary department dedicated to education and research in all areas of modeling, virtual environments, and simulation. See <http://www.movesinstitute.org>.

Fortunately, the costs of most such efforts to promote creative practices tend to be modest. At the low end (e.g., to support a seminar series), the incremental costs might involve little more than refreshments, photocopying, and a quarter-time graduate student assistant; to provide the well-stocked high-tech site described above would, though, involve hardware, software, networking, facilities, and technical support costs.

## SERVICE UNITS

As is true across society, academic environments depend on IT tools and infrastructure for their day-to-day activities in general and also for the support of ITCP work specifically. As such, supporting services are valuable—indeed essential—to the work of today's university. For this discussion, though, the question of interest is how (or whether) such supporting services can directly affect the rise of new intellectual activity aimed at understanding what IT itself could become and how it can be integrated with elements from the arts and design world to represent and explore ideas in different ways. Whether the assistance takes the form of consultation with a database software expert in a centralized IT department or is provided through a service course (e.g., "Introduction to Programming for Humanists"), the support provided by service units does not usually lead directly to deep insight into IT—how IT is developed, how it works, how it could be meaningfully incorporated into other work, and how it could be extended in support of new explorations and objectives. Mastery of a new set of IT-based tools[16] is only the first step for artists and designers involved in such efforts, as it has been for scientists. It is not clear that there is in the arts and design an analogue to the IT service unit, other than the selective use of artists or designers, who tend to be incorporated as needed into technical projects rather than centralized in a service unit.

Service organizations can be important in stimulating ITCP work in environments where such explorations are otherwise difficult to launch. In addition to the classical IT service unit, a library—already a nexus of multiple disciplines that embodies IT in its daily practices—could help foster IT's introduction into arts and design-oriented departments, especially in institutions without engineering schools or computational science programs. The transformation of both libraries and the schools that prepare library professionals illustrates a fundamental attention to IT as a tool for and means of cross-disciplinary exploration.[17] For example, specialized digital libraries such as the Perseus Digital Library at Tufts University can simplify access to disparate literatures and thus help enable work based on multiple disciplines.[18] Physical spaces are also evolving, such as Vassar's Me-

*Service organizations can be important in stimulating ITCP work in environments where such explorations are otherwise difficult to launch.*

---

[16]Understanding how information technology works has been called IT fluency. See Computer Science and Telecommunications Board, National Research Council, 1999, *Being Fluent with Information Technology,* National Academy Press, Washington, D.C.

[17]Two institutions of higher education that were previously more narrowly focused on library science and now have a much broader, social science and policy-rich curricula are the University of Michigan's School of Information and the University of California at Berkeley's School of Information Management and Systems. See <http://www.si.umich.edu> and <http://www.sims.berkeley.edu/>.

[18]See <http://www.perseus.tufts.edu/>. A general resource on collaborative facilities for academic environments is under development as a project that is hosted by Dartmouth College and sponsored by the Coalition for Networked Information. See <http://www.dartmouth.edu/~collab>.

dia Cloisters, a collaborative space in the college's main library that is equipped with state-of-the-art technology,[19] and the Media Union at the University of Michigan, which brings together information resources, information technology, production studios, and the combined talents of information professionals from across campus units to facilitate cross-disciplinary collaboration as well as integrative learning and exploration.[20]

●  ●  ●  ●  ●  ●  ●  ●  ●  ●  ●  ●  ●  ●  ●  ●  ●  ●  ●  ●  ●  ●  ●  ●  ●  ●  ●  ●  ●  ●  ●  ●  ●  ●  ●

# FOSTERING ITCP WORK WITHIN MAINSTREAM DEPARTMENTS AND DISCIPLINES

## COMPUTER SCIENCE

Computer science is a young discipline (departments and degrees in the field emerged in the 1960s) with roots in mathematics and electrical engineering. Although it drew from other disciplines, and as what Herbert Simon called an "artificial science"[21] would seem to have inherent flexibility, its recent history of cross-disciplinary interaction—that is, in the period when ITCP has been emerging—has been mixed. In the late 1980s and early 1990s, for example, attention to applications of computer science was sometimes questioned in the computer science community, in part because of concern about the potential for diversion of resources to other fields using computing.[22] As applications have proliferated, respect by the field for research in computer science has most often been accorded to work defined as contributing to advances in the science and engineering that are fundamental to the field itself, whether or not an application problem has inspired the activity. Collaboration with, or engagement of, any field is judged within computer science by that criterion—which is a reflection of the basic view that good work in the field advances the field.[23]

---

[19]See <http://mediacloisters.vassar.edu>.

[20]See <http://www.ummu.umich.edu/intro.html>.

[21]Herbert A. Simon, 1996, *The Sciences of the Artificial*, Third Ed., MIT Press, Cambridge, Mass.

[22]Although CSTB's 1992 report *Computing the Future: A Broader Agenda for Computer Science and Engineering* (Computer Science and Telecommunications Board, National Research Council, Juris Hartmanis and Herbert Lin, eds., National Academy Press, Washington, D.C.) had as a first recommendation sustaining the core of the field, its recommendation that the field look to application areas for inspiration was controversial when published.

[23]For example, collaborations with psychologists and sociologists have been routine in work on usability and human-computer interaction, including in particular the area of computer-supported cooperative work. Because research in these areas involves studying how people use technology and draws on those insights to improve the design of hardware and software, such collaborations have had some direct appeal, as reflected in professional society, conference, and journal activities and in the composition of research teams.

This somewhat limited focus, however, is not conducive to the exploration of new relationships called for in ITCP work.

## Examples of ITCP Work

There are computer science faculty members pursuing ITCP work,[24] although the number is relatively low, in the committee's estimation, compared to the number of art and design faculty who pursue ITCP work.[25] For example, Ken Goldberg, a professor of industrial engineering and operations research at the University of California at Berkeley, explores ITCP through robotics, including telerobotics (operated over the Internet), which he has extended into telepistemology (see Box 6.2 for a description of his work). For about 15 years he has been "using art work as a way of challenging—questioning, critiquing—this world that I was operating in, the world of robotics and engineering." His work has attempted to explore the human and philosophical side of issues in control technologies. He observed that the engineer/researcher in him wants to discuss operations, steps, and motives very precisely, whereas the artist in him is both skeptical about and has an appreciation for the benefits of saying less.

As described in Chapter 2, Michael Mateas, a new faculty member at the Georgia Institute of Technology, uses the term "expressive AI" for his combination of art studio practice and artificial intelligence (AI) research practice; he uses AI to build interactivity into cultural artifacts, applying technology to enable a "negotiation of meaning between the artist and the audience." In describing his work to the committee,[26] Mateas sketched out the principal schools of thought in AI and then differentiated what he did from them, explaining differences in the goals of art or cultural production and AI research and in the ability to shape the internal workings of a device as they relate to the audience experience.[27] In sum, "there is new AI technology being invented that would not be invented or discovered or even thought about by someone doing AI research who's not interested in art. But it is focused on art practice."

---

[24]They include David Salesin at the University of Washington and Roger Dannenberg at Carnegie Mellon University, members of the committee. Also see the program of the conference MOSAIC 2000, available online at <http://www.cs.washington.edu/mosaic2000/program.html>.

[25]The committee does not have specific quantitative evidence to support this claim, which reflects the committee's expert judgment, based on professional networking by its members.

[26]In January 2001, when he was a doctoral student at Carnegie Mellon University.

[27]Conventional AI, as with other computer science, focuses on quick and efficient completion of a task, whereas cultural production focuses on poetics and rich possibilities for interpretation by an audience; AI is concerned with generalities (principles), whereas cultural production focuses on the potential of a specific work, per se; and AI aims at realism while cultural production aims at abstraction, which may exaggerate or change a part of the world. Also see Box 2.3.

## BOX 6.2
### Telerobotics

The study of knowledge acquired at a distance, telepistemology, tries to comprehend how epistemology can inform the understanding of robotics, and to what extent telerobots can furnish new insight into classical questions about the nature and possibility of knowledge. Ken Goldberg, the editor of *The Robot in the Garden*,[1] sees a fine line between how people perceive traditional virtual reality, where interactions take place in synthetic space, and telematic reality, where there is a physical space but it is distant. How people interact in telematic reality is a central concern of telepistemology, and it raises important technical and moral questions. Technical questions include whether telerobotics provides access to real knowledge and the degree to which telerobotic experiences are equivalent to proximal experiences. Moral questions concern how people should act in telerobotically mediated environments and to what degree technological mediation affects human values.

One of Goldberg's telerobotic art projects, Telegarden (Figure 6.2.1), tries to provide insight into some of these questions by focusing on how information arises from live interaction with remote physical environments. Goldberg describes Telegarden, which has been housed at the Ars Electronica Center in Austria since 1995, as a telerobotic art installation on the Internet where remote users direct a robot to plant and water seeds in a real garden. The Telegarden receives, on average, more than 1500 hits a day.[2] Based on the feedback received from visitors to the Telegarden site, Goldberg and his colleagues have been able to reflect on how perception, knowledge, and agency, important principles of telepistemology, are being defined in telerobotic interaction. For example, they have found that people are skeptical of the experience, often wondering whether the Telegarden is real in the physical sense or a digital simulacrum.[3]

FIGURE 6.2.1 Telegarden. Image courtesy of Ken Goldberg.

---

[1]Ken Goldberg, ed., 2001, *The Robot in the Garden: Telerobotics and Telepistemology in the Age of the Internet*, MIT Press, Cambridge, Mass., and London.

[2]At its peak, the site attracted on the order of 15,000 hits per day.

[3]Ken Goldberg briefed the committee at its meeting in January 2001 at Stanford University. For further information concerning his work, see <http://www.ieor.berkeley.edu/~goldberg>.

Different institutions may be culturally better suited to different models of ITCP work. For example, committee members' experiences, and testimony to the committee, suggest that Carnegie Mellon University (CMU) generates cross-disciplinary activities in various forms with relative frequency and success as compared to many other universities. In the CMU culture, it is generally acceptable to work in any discipline from a basis in any other, an unusually open posture (albeit bounded by expectations for performance that can be evaluated within the home discipline[28]). Work in different disciplines is accepted, too: There is a separate history of computer scientists with academic appointments who are also significant creative artists, namely composers. One can speculate that such a development took place because composers were more likely to program computers than were other artists or designers.

Regardless of setting, a key element of success is effective training; more than talking about opportunities is needed. In the early days of a new area of study, people put in a lot of time to educate themselves in the new domain. Time matters, and accordingly, so does institutional support for the investment of time necessary for learning.

What goes on within the walls of academia is only part of the ITCP story, however. Computer science laboratories and departments have both official programs of activities and a panoply of unofficial activities—a penumbra—that extends the interface between the academic and the non-academic, both feeding and drawing from popular interests off campus. Comparisons might be drawn to the intellectual fringe around industries that deal with pop culture, or the marginal art production feeding off the core, which serves as a magnet. It is important to sustain that penumbra (supporting such activities as short-animation/student festivals, for example) as it relates to part of a strategy of nurturing academic ITCP on the computer science side. This is the arena in which collaborations with independent artists may be achieved.

Among the most notable of such interactions—and the most visible—were the computer graphics and computer music collaborations among computer scientists and artists and designers dating back to the early years of the computer science field. Computer graphics evolved in part through interactions with scientific (and other) users whose needs for visualizing complex processes could be met with this technology.[29] See Box 6.3.

---

[28]Also note the further comments on criteria for tenure in the section "Challenges in Computer Science Departments," below.

[29]Scientific visualization has itself evolved with computational power in ways that reinforce the values of an artistic approach to making sense of growing quantities of data. Thus, for example, artist Donna Cox was an early employee of the National Center for Supercomputing Applications at the University of Illinois at Urbana-Champaign, where she has collaborated with computer scientists and other scientists on the challenge of visualizing the processes for which supercomputing generates enormous volumes of data.

**BOX 6.3**

**Beyond Academia: Computer Science Reaches Out to the Community**

The development of computer music and computer graphics collaborations among computer scientists and artists dates back to the early years of computer science. Computer graphics emerged from computer science research in the 1960s, spurred by vigorous Defense Advanced Research Projects Agency (DARPA; then ARPA) funding,[1] but it also grew up in a variety of pockets of filmmakers working in close association with both academic and industry-based laboratories. Computer music also took a variety of paths, linked to corporate research programs as well as university computer centers.

Much of the foundational work in computer-generated sound occurred at Bell Labs, under the leadership of Max Mathews and John Pierce, who from the earliest days employed qualified composers to collaborate in the study of both the acoustics of instruments and the design of high-level languages for composition.[2] Links to computer science departments were highly visible as well. Stanford University's Center for Computer Research in Music and Acoustics, which leverages computer technology as an artistic medium and research tool,[3] grew out of the Stanford Artificial Intelligence Lab, and centers for hybrid analog-digital computer music were located at Princeton University, the University of Toronto, and Columbia University in the 1960s. (The formal properties of avant-garde music were well suited to procedural representation even earlier than this, as exemplified by the work of Xenakis in the 1950s, who worked with IBM researchers in France to devise stochastic composition under computer control.[4]) As smaller systems came within reach of independent composers and more academic music departments in the 1970s, composers themselves tended to take the lead in devising both hardware and software components. By the 1980s, easily programmable commodity hardware became affordable, and a communication protocol—the musical instrument digital interface (MIDI)—made it possible for both researchers and commercial practitioners to adapt the cumulative know-how in languages and synthesis techniques across a broad spectrum of musical styles. Indeed, an unanticipated consequence of this widely adopted protocol was an important factor in enabling the growth of new-media art forms, such as interactive installation and performance art: The musical event-generator could as easily talk to image-making, lighting, motion-detection, or robotic machinery as to synthesizers or sound-processing devices.

## Challenges in Computer Science Departments

One of the biggest challenges to be addressed in departments of computer science seems to be a culture that discourages ITCP work—although sometimes subtly. Committee members and briefers to the committee recounted personal stories of their art and design interests generally not being viewed as productive. As one individual explained, "A faculty member told me that it was a complete waste of time and that I should stop doing the 'flaky' stuff. That caused me to go underground." This person decided to limit conversations with other technical colleagues about artistic work and did not include the work in the case for promotion, observing that relevant activities such as seminars were treated as over and above conventional responsibilities. Another briefer characterized the department's reaction to art and design work as "friendly toleration."

In computer graphics, the technical issues addressed in the 1960s by the DARPA contractors (Massachusetts Institute of Technology, Stanford University, Carnegie Mellon University, University of Utah) dealt primarily with the problems of three-dimensional modeling and simulation—fueling the decades-long quest for the Holy Grail of immersive photo-realistic virtual environments.[5] Important contributions were made to the commercial motion picture industry by university laboratories in the 1970s—notably the New York Institute of Technology—which sustained the core set of early DARPA researchers before price and performance factors leaped forward in the 1980s and made a market for commercial graphics a reality. The National Film Board (NFB) of Canada collaborated with the Canadian National Research Council in producing some of the most widely viewed and convincing artistic computer animations before 1975. This collaboration was decisive for the NFB's early adoption of digital techniques, and it fostered a community of advanced users with strong links to the academic work in Canadian labs (at the Universities of Waterloo, Toronto, and Montréal) that helped to underwrite the knowledge base feeding the successful cluster of Canadian graphics software firms—Alias, SoftImage, Sideffects, and Discrete Logic.

The scientific and entertainment graphics community coalesced around the professional organization SIGGRAPH (the Special Interest Group on Computer Graphics and Interactive Techniques of the Association for Computing Machinery), which has been successively extending its scope to include other scientific specialties, such as human-computer interaction, and media genres, such as computer games. It is perceived as emphasizing commercial applications, which have games as a driver for computer graphics.

---

[1]See Computer Science and Telecommunications Board, National Research Council, 1999, *Funding a Revolution: Government Support for Computing Research*, National Academy Press, Washington, D.C.

[2]Pierce worked with Claude Shannon on pulse-code modulation, and Mathews followed up with the analysis of transmission systems using picture-processing algorithms. They had an abiding interest in audio and visual; Mathews worked closely with Lillian Schwartz, a computer artist at Bell Labs beginning in the late 1960s who was also associated there with A. Michael Noll and Ken Knowlton, who pioneered certain graphics and animation programs.

[3]See <http://ccrma-www.stanford.edu/>.

[4]Iannis Xenakis, 1992, *Formalized Music*, Pendragon Press, Stuyvesant, N.Y. (first published in 1963 as *Musiques Formelles*, Editions Richard-Masse, Paris).

[5]See the chapter "Virtual Reality Comes of Age" in Computer Science and Telecommunications Board, 1999, *Funding a Revolution*.

---

The criteria for tenure in computer science tend to reinforce such negative attitudes toward ITCP work. Although they vary among computer science departments, criteria for tenure at a research university typically include the ability to win a strong endorsement in confidential assessments by subject experts at other institutions and the quality and quantity of published research, as well as the quality of the candidate's research program, supervision of doctoral students, and general teaching record. If the subject experts on a tenure committee do not include anyone familiar with ITCP work, or if that ITCP work does not produce a bounty of published research in major technical journals, efforts devoted to ITCP work may not advance a computer scientist's career within the department.

Another challenge is related to the rise of computing and data communications across the board in academic environments, with elements of computer science increasingly being taught in other departments—a situation that makes some computer scientists skeptical

One of the

biggest

challenges to be

addressed in

departments of

computer science

seems to be a

culture that

discourages ITCP

work—although

sometimes subtly.

about collaborating with colleagues who may be viewed as lacking the credentials to be hired in the computer science department itself. When computer science is taught as a service course to meet specific needs, other departments may be less motivated to reach out to computer scientists for help with deeper, challenging problems in computing and communications. The computer science diaspora in academia has typically focused on developing skills of immediate application in another field, whether programming (at the more sophisticated level) or the use of common software tools (at the less generic user level). Whether they are expected to teach computer science (CS) courses for majors in other departments or whether those departments internalize such courses, computer scientists tend to see such activity as something less than "doing real computer science." Also, the emphasis of such courses on meeting immediate needs and delivering a targeted service is not conducive to thinking about collaborations.

The willingness and ability of academic computer scientists to engage with artists and designers depends on institutional support, as noted above, and on other particulars of a given department—whether, for example, the department is in a school of engineering, whether a bachelor of arts degree is an option, and so on. Recently, to improve the richness of its cross-disciplinary interactions, the CS department at MIT planned to make about half of its new appointments joint with another unit at MIT:[30] One course per year would be taught in each department, salary and promotion decisions would be made jointly, and the research expected from the new faculty member would appeal to each department, with selection standards the same as for core (non-joint) faculty. The first person hired in the CS department under this plan was a physical chemist, and the partnering department was bioengineering. It was expected that he would be exposed to new methods in computer science and in turn would contribute to new problems demanding new CS solutions—and thus motivating fruitful cross-disciplinary interaction. How this department's open-ended plan for cross-fertilization is realized and whether it will contribute to ITCP will depend, in part, on whether artists (for instance) can also be hired. That prospect is uncertain, in part because of questions about how to assess candidates who are so different from computer scientists and how to compensate people who span departments with substantially different salary structures.[31]

Whatever the mechanism for forming collaborations for ITCP across departments and disciplines, they must include researchers working with shared goals and must address intellectually challeng-

---

[30]Briefing to the committee by John Guttag, chair of the Electrical Engineering and Computer Science Department at MIT, May 31, 2001.

[31]On the other hand, the differences might make artists and humanists more attractive than social scientists, who bear the "soft science" stigma among some computer scientists because of perceived contrasts in the quantitative and analytical elements of the fields. At MIT, the point may be moot, because the Media Lab already provides a vehicle for exploring ITCP.

ing computer science problems. One approach to ascertaining whether a project is intellectually challenging might be to ask whether it could lead to a suitable computer science doctoral thesis. Computer science faculty tend to value work that advances the state of the art in computer science. For dual appointments, the issue may be whether the work helps to redefine computer science for the future or to create a new field.

One approach to establishing credibility for, and perhaps encouraging, ITCP work and programs within computer science departments would be to appoint a senior researcher as the leader of cross-disciplinary activities. Sending such a signal may be even more important for collaborations that would include a field traditionally regarded as far from computer science. In addition, institutional paths are needed that allow faculty members to break out of narrow specializations and that encourage cross-disciplinary Ph.D. programs, although market realities may argue for a "home" department in computer science or another technical area. Programs with cross-disciplinary minors could also be established or expanded. Cross-departmental activities such as distinguished lecture series and topical study groups or seminars could also be adopted.

A key criterion in assessing specific options for cross-disciplinary appointments will be their potential to advance the interests of graduate students, who are the backbone of a computer science department's research program. It will be necessary to frame interesting new, cross-disciplinary problems that provide appropriate entrées to careers for graduate students (perhaps more at the master's than at the doctoral level).

## ART PRACTICE AND DESIGN

Within the academic environment, the arts range broadly from traditional fine arts—the visual arts, music, drama, dance, photography, film and video, and so on—to design, architecture, crafts, and the history and criticism of the arts.[32] The rise of IT has touched all of these areas, some more than others, blurring some of the traditional distinctions among them. The emphasis of arts programs varies with the institution: Some take a more humanities-based approach, teaching artistic expression and critical theory as a way of understanding cultural problems, or a more theoretical approach, studying the fine arts' interrelationships with all aspects of contemporary culture; others prepare students for careers in commercial graphics design, fashion, or animation and special effects. And some academic art environments have a broader range of exposure to IT than others.

> One approach to establishing credibility for ITCP work and programs within computer science departments would be to appoint a senior researcher as the leader of cross-disciplinary activities.

---

[32]The rise of IT in the 1990s is viewed as an important cultural phenomenon with far-reaching implications for visual artists who wish to interact with, and represent, the world in which they live.

The outreach from the arts to information technology, like that from the humanities,[33] echoes the sciences' earlier reaching out to IT. Computational science has been dominated by people and activity based on one or another science, although new, hybrid programs—academic programs, industrial and academic laboratories—have emerged. Within the field of computer science, computational science has tended to be viewed as an outside(rs') activity, in part because the majority of the impetus came from the various sciences that saw in computer science useful tools. This is part of the previously noted computer science diaspora. Response from within computer science required recognition that the field could gain inspiration from these other fields, rather than serve only the one-way function of supplying those fields. Similar concerns and patterns shape academic ITCP, still embryonic. Further complicating the picture is the fact that ITCP may involve more than the pairing of arts with computer science (or electrical engineering). This is illustrated in particular by the recent trend to involve as well ideas and activities from the life sciences, given growing use of IT in that domain and the visual or music performance potential that results.[34]

Some arts programs have embraced IT, even creating independent areas of study; many have simply broadened their current programs by adding courses in various software packages, thereby treating IT as another tool. Arts programs are increasingly encouraging intermedia and cross-disciplinary artwork as IT tools blur the distinctions between traditional art disciplines. Some art schools are more skeptical of IT's potential as "new media" and argue that computers and software, input devices such as digital cameras and scanners, and displays such as LED screens and projectors are nothing more than tools that replicate and simulate traditional media without presenting any new problems in visual theory.

> **Arts programs are increasingly encouraging intermedia and cross-disciplinary artwork as IT tools blur the distinctions between traditional art disciplines.**

---

[33]See Computer Science and Telecommunications Board, 1997, "Computing and the Humanities: Summary of a Roundtable Meeting," ACLS Occasional Paper 41, American Council of Learned Societies, New York, N.Y., available online at <http://www.acls.org/op41-toc.htm>.

[34]Take the case of the artist as creator of new life forms. Eduardo Kac, of fluorescent green-bunny fame (see his Web site at <http://www.ekac.org>; see also Eduardo Kac, 1998, "Transgenic Art," *Leonardo Electronic Almanac* 6(11): 289-296), in another project, biblically titled "Genesis," mutated *E. coli* bacteria into genesis bacteria. What is of specific interest from the perspective of IT is that the genomic code was translated into Morse code, which was then converted into DNA base pairs. The complicated production of this synthetic gene is not the concern here. What is significant, however, is that these cultured fluorescent bacteria remain invisible until IT intervenes. A microscope with an internal ultraviolet (UV) light source, a Web server, and an over-life-size video projection make visible the separation and interaction processes of these newly created entities. Both the onsite and the online observer can intervene in the procedure of bacteria transformation by manipulating the UV light source. A further aspect of this installation (shown at the Ars Electronica conference in 1999) is the DNA-Music-Synthesis—turning the physiology of the DNA into a musical parameter. The real as well as the online viewer is influential here, too, since she can control the rate of mutation and the sequencing.

## SCHOOLS OF ART AND DESIGN

To understand the emergence of ITCP from the academic arts perspective, it is helpful to consider how major schools of art and design approach IT. The Rhode Island School of Design (RISD), for example, observes that "computers are natural allies in the creative process, with the emphasis on using new technologies to express artistic vision. Breakthroughs in digital media have transformed virtually every discipline taught here."[35] Indeed, IT has influenced the development of RISD programs in graphic design, apparel, ceramics, architecture, photography, film, animation, video, and illustration—providing new ways to address a shared concern in visual literacy and creative problem solving.

RISD's integration of digital media into its more traditional course work follows the typical pattern of the integration of IT into arts education—the first and most natural step in the process is to use software to create a model of a potential artwork before the work is realized in a traditional medium, such as paint or marble. Applications such as Adobe PhotoShop, and any number of three-dimensional modeling applications, allow students to experiment with image composites and collage, placement of colors, and general composition before working with physical materials. However, in 2002, RISD announced a new graduate department to explore innovative approaches to digital media. Expanding on a media art focus, the vision of the new department is to provide a diverse environment for multidisciplinary and transdisciplinary exploration of digital media.

Schools such as the California Institute of the Arts (CalArts[36]) have focused on preparing undergraduates for professional careers in art, design, film, music, and theater. Information technology has been wholeheartedly incorporated into all of the above programs, and a new, cross-disciplinary study option has been created. Graduate students have the option of undertaking a new concentration in Integrated Media, designed for students who are fluent with (or want to be fluent with) computer programming and digital technologies as basic elements in their artwork. Programming is taught to increase mastery of IT (beyond learning how to use an existing package).

New York's School of Visual Art (SVA) boasts an impressive list of technical courses in its M.F.A. program in computer art, including C++ programming, UNIX, interactive multimedia, interface design, and sound—a list that shows a range from key skills development (e.g., programming) to explorations in applying those skills. The SVA's computer art M.F.A. was the first of its kind, established in 1987.[37]

---

[35]RISD's Web site even features a technology statement. See <http://www.risd.edu/technology.cfm>.

[36]See <http://www.calarts.edu>.

[37]Information on the School of Visual Arts is available online at <http://www.sva.edu/mfacad/facilities.html>.

A course called "Public Art/Private Spaces" in CMU's College of Fine Arts explores IT's effect on the notion of public space, and it directs students toward the creation of a public artwork that addresses these issues. "With the ubiquity of advertising, cell phones, surveillance cameras and wireless computers it is increasingly difficult to define the parameters of public space. In this course, students will consider how visual artwork and performance are incorporated into what is considered 'public space.'"[38] Carnegie Mellon's M.F.A. program is rooted in critical theory and a conceptual approach to art making, and consequently has incorporated IT into its broader cultural concerns as well as into the practical aspects of streamlining the art-making process and working in new media. Also see Box 6.4 for an example of a program in electronic literature.

The Department of the Arts at Rensselaer Polytechnic Institute (RPI) provides a contrasting model to CMU, as it operates as the sole creative arts department within the School of Humanities and Social Sciences. Stressing "integrated electronic arts" in the wider context of a technological university, the Arts Department at RPI offers both a concentrated graduate master's degree in electronic arts practice and an undergraduate dual degree in electronic media, art, and communication. Without the richness of creative resources and disciplinary specialties found in colleges or faculties of fine arts, RPI's faculty of primarily electronic artists must attempt to cater to both the high end of experimental, research-oriented graduate ITCP practice, as well as a large and ever growing undergraduate demand for a hybrid bachelor's degree. After experiencing rapid growth in the late 1990s, the RPI Arts Department is now poised to become the anchor academic partner for the new Experimental Media and Performing Arts Center, which will provide exhibition and performance venues for both locally developed and internationally curated ITCP work.

The arts and design departments and colleges have generally embraced information technology as a new tool to accomplish work. However, somewhat fewer programs in the practice of arts and design have focused on ITCP work—to examine how IT may be able to enable new forms of art and design and new ways of making such work available, as well as how IT may be able to change profoundly the practice within traditional art and design categories. In some respects, art and design departments were challenged just as society at large to identify how advances in IT could be used beneficially. In one sense, expectations are higher for arts and design departments than for computer science departments: Most people would readily conclude that some kind of profound development in the arts and design should emerge from advanced IT capability. By contrast, it is somewhat less clear how ITCP could and should affect computer science departments and whether the nature of this influence should be minor or major.[39]

---

[38]See <http://www-art.cfa.cmu.edu/academic/descriptions.html>.

[39]The committee argues for a major impact—see the discussion in Chapter 4.

---

**BOX 6.4**

**School of Literature, Communication, and Culture at the**

**Georgia Institute of Technology**

About 15 years ago, the English Department of the Georgia Institute of Technology (Georgia Tech) received a large institutional grant from the National Endowment for the Humanities to transform itself. The department broadened its scope to include electronic media, leading to the adoption of a new name: the School of Literature, Communication, and Culture (LCC). Aggressive recruitment for media arts faculty followed, and with a number of new members secured, the school began to focus on growth. First an M.A. program in information design and technology was created. Then a series of workshops were convened at corporate sites, which proved to be an excellent financial resource as the workshops soon raised more than $1 million in discretionary funds. This revenue went toward the purchase of up-to-date equipment and to fund traditional courses, research, and projects in literary studies. Today, the LCC is widely recognized as one of the premier programs in electronic literature.

The transformation of a traditional English Department into a non-traditional electronic media school has resulted in both advantages and obstacles. The success and prestige that the school enjoys within Georgia Tech and the new-media community at large have created visibility that would have been unavailable if it had remained a traditional English department. With the help of creative campaigning by the LCC, the engineers and scientists of Georgia Tech, faculty members who hold the majority of clout within the university, have had little trouble understanding the benefits and advantages of a program in digital media. The school offers two graduate programs, continuing education programs, and an undergraduate major, two minors, and three certificate programs.[1]

The changes also have created tensions. The new focus involves commercial considerations to a much greater extent than one finds in a traditional English department—perhaps welcomed by some faculty, but troubling to others. Related to this tension is the concern that the department might be morphing into a fancy kind of trade school, away from an established place in the world of scholarship—in other words, that teaching highly marketable electronic skills might come at the expense of intellectual and scholarly inquiry. One solution may be to approach electronic literacy as a rhetorical practice, rather than a skill.

---

[1]See the LCC mission statement at <http://www.lcc.gatech.edu/index.html>.

---

As departments within universities, art practice and design programs must adhere to the same general guidelines for tenure as computer science departments. However, there are some important differences attributable to the discipline and profession. Perhaps the central difference involves what constitutes a publication. In both cases, completed work must be made public in some way. Computer science emphasizes the publication of articles and technical reports (and books to a lesser extent). Publication in the arts and design may occur through exhibition, performance, or other means; there can be a scholarly publication associated with a completed work, but not necessarily (and in computer science, there may be a prototype or working system that corresponds to a scholarly article, but not necessarily).[40]

---

[40]Experimental computer science focuses on demonstrations, somewhat analogous to exhibition or performance in the arts and design world, in which the primary focus is on systems that operate, not on articles describing systems.

# CROSS-CUTTING ISSUES

Academic organizations can be—and often are—incubators for new fields that emerge from the interstices of the old ones. That was the case for computer science—and for the study of film and video, for example. Both of those arenas, however, illustrate how emerging fields of study have to struggle to gain respect in the academic community. This struggle implies that at early stages, resources and institutional policies can constrain potential.

## HIRING FACULTY

At first glance, co-hires (appointments with nearly 50 percent in each department—in computer science and the arts and design) may seem to be an attractive prospect. The co-hire becomes steeped in diverse settings related to ITCP and can draw on a tremendous range of expertise and resources. However, entering into a joint appointment or major collaboration across disciplines introduces challenges and risk, especially for junior faculty who have yet to attain tenure, inasmuch as decisions about their promotion and tenure may depend on assessments by people accustomed to more traditional avenues. The demands of the tenure process may be one reason that venturing into new and cross-disciplinary areas seems more suitable to senior faculty[41]—but it is often the young who show more interest in new areas, especially those involving IT in a new context.

Thus, the committee believes that faculty should have their primary home in one department. This home department's criteria are used for tenure and promotion decisions. The faculty member would attend the home department's faculty meetings and seek mentorship from the ranks of this department. In some instances, a minority appointment in a second department can be desirable to encourage cross-departmental ITCP research and teaching (which may result, for example, in teaching a course in the second department occasionally, perhaps jointly with another faculty member whose primary appointment is in the second department). The utility of zero-percent appointments[42] is questionable, especially if the reality is that the faculty member seldom becomes involved in any way with the second department.

> At first glance, co-hires in computer science and the arts and design may seem to be an attractive prospect.

---

[41]With a focus on computer science, this observation was discussed in CSTB's *Making IT Better: Expanding Information Technology Research to Meet Society's Needs* (Computer Science and Telecommunications Board, National Research Council, 2000, National Academy Press, Washington, D.C.).

[42]Zero-percent appointments provide faculty members with formal recognition and selected privileges (which may include attending and voting at faculty meetings, the opportunity to supervise doctoral students, and so on) within a department, even though that department does not provide any of the faculty member's salary.

One alternative is to create voluntary clusters of interested people within an art, design, or computer science department who would teach material on the edge of the respective department. For example, there could be an art/IT group within an art department. It is important to locate IT quasi-formally within all fields simultaneously—even if it is only at a voluntary cluster level at first. This parallel existence would serve to familiarize larger groups of faculty and students with ITCP possibilities and help to increase the awareness of ITCP work. The idea of voluntary clusters could also be enabled and bolstered by the workshop concept (see the section "Workshops" above).

Some programs in ITCP areas depend on adjunct faculty. As in other domains, the limited use of adjunct faculty can be quite beneficial in bringing different perspectives (often from areas in the relevant practitioner community) into academia. However, dependence on adjunct faculty in ITCP centers as a cost-savings measure is a flawed long-run strategy, although budget pragmatics may offer no alternative in the short-run in some instances. A large cadre of adjunct faculty jeopardizes sufficient stability among the faculty to develop the deeper insights that can derive from projects that extend beyond one term or one academic year. Furthermore, a large cadre of adjunct faculty promotes the idea of second-class citizenship in the academic community for those who pursue ITCP work.

> A large cadre of adjunct faculty promotes the idea of second-class citizenship in the academic community for those who pursue ITCP work.

## ENCOURAGING MULTISKILLED INDIVIDUALS AND COLLABORATIONS

Academic environments feature metamorphosis in core disciplines—computer science, various arts and design—and more or less independent cross-disciplinary activities, which aim to generate multiskilled people and support their work. Both individuals with skills from multiple disciplines and people willing and able to collaborate across disciplines appear important for sustaining cross-disciplinary programs.[43] Suitable training and institutional support for either category are still emerging.

A challenge is to overcome the tendency of the communities within the ITCP realm to fear becoming marginalized in collaborations, to fear being treated as technicians in support of another's work. Ensuring sharing of responsibility and recognition is prerequisite for collaboration and for acquisition of cross-disciplinary skills. That challenge may be greatest on the computer science side, from which ITCP has to be able to draw to achieve its full potential.

Because the best collaborations seem to be modest-sized efforts that grow from the bottom up (graduate students often may be the

---

[43]See Veronica Boix Mansilla, Dan Dillon, and Kaley Middlebrooks, 2002, "Building Bridges Across Disciplines: Organizational and Individual Qualities of Exemplary Interdisciplinary Work," review draft, Project Zero, Harvard Graduate School of Education, Cambridge, Mass.

ones to make and promote connections across boundaries because they are not yet fully invested in the models and norms of a particular field), broader exposure and effective evangelism may be needed to overcome the disciplinary inertia that can militate against interactions with people who seem very different. Administrators need to recognize that effective cross-disciplinary activity is harder and can involve more work than traditional intradisciplinary efforts.[44] Accordingly, they should provide incentives for people from different disciplines to work together and should make it easy to get started by minimizing the red tape, while maintaining perspective on academic goals and on how best to structure the means toward those ends. Investments are needed in planning and preparation and in guidance for students. For example, the tendency for fellowships to be attached to departments may limit flexibility. Further complications stem from differences across disciplines and departments in compensation, teaching load, graduate student culture (e.g., autonomy in dissertation selection, expectations for authorship credit on published work, and career expectations beyond the terminal degree), and support (i.e., it is not unusual for computer science graduate students to receive research assistantships during much of their course of study, but this is much less common in the arts and design fields). Although these differences can be found in comparisons between any pairs of fields, they are greater in cases involving major historical, cultural, and resource disparities, such as those between the arts and design on the one hand and information technology and computer science on the other (see Chapter 8 for further discussion).[45]

Some universities such as Carnegie Mellon can draw on strengths in both computer science and art, both of which attract a critical mass of talent from which interactions may flow. The case of CMU illustrates the benefits that can accrue from large scale, which is a circumstance that may not be broadly transferable. For example, the Human-Computer Interaction Institute at CMU has organized a series of lunches engaging people from the computer science and design departments, and a recent course drew from both the robotics and art departments to explore robotic art. Few institutions could bring together on a regular basis audiences with such a wealth of expertise

---

[44]See Catherine Sentman and Samuel Hope, 1994, "Disciplines in Combination: Interdisciplinary, Multidisciplinary, and Other Collaborative Programs of Study," briefing paper of the Council of Arts Accrediting Associations, Reston, Va., March.

[45]An imbalance in resources can have a significant effect on the nature of collaborations pursued. Suppose that two professors—one in the arts and one in computer science—wish to pursue a joint project. Further suppose that the project will require hardware and software that cost $150,000 and four half-time research assistants over the course of several years. To the extent that a resource imbalance exists, such projects may have to be funded primarily by the computer science participant (introducing a power asymmetry into the collaboration) or not be pursued.

and experience, but those able to do it have a significant advantage in the support of ITCP work.

Administrators also have to resist the urge to direct large sums of money to anything digital. A new organization, for example, is not necessarily the best solution, at least in the beginning. And they also need to beware of the faddism that can accompany identification of cross-disciplinary opportunities. In particular, university administrators may feel tension about their institutions being perceived as (or becoming) fancy trade schools if programs lack a recognized research base.

However, in the long run, the creation of a specialized center is desirable to consolidate common interests as demonstrated by the initiatives of individual faculty members. ITCP work must develop its own world view, and while that will necessarily be built largely from the parent disciplines, a more neutral context is required so that the inherent problems and contradictions can be negotiated.

## DESIGNING CURRICULA

Questions arise about kinds of degree programs and how they should relate to skills. There is an art to evolving courses—in computer science and in the arts and humanities—to promote cross-fertilization, and it is not clear that there are replicable models. Different approaches will be needed where the emphasis is on increasing breadth versus depth of knowledge. On the arts side, a challenge is to move beyond treating IT as black-box tools and to enable more exploration of the workings of IT, to foster deeper integration of art and IT. On the computer science side, a big constraint is the already tight engineering curriculum that shapes most degree programs (though most programs include some social science and humanities courses as a part of a general education requirement). The challenges begin with delivering arts and humanities (or social science) courses in ways that make clear the relevance to computer science students, and computer science courses in ways that link other issues and contexts, thereby opening the doors to cross-disciplinary inquiry. The next level of challenge pertains to curriculum. As Sentman and Hope ask,

> Is there a common, clearly defined concept of what students should know and be able to do after completing a program that combines work in two or more disciplines? How do collaborative programs aid in the development of knowledge and skills both in the component disciplines and in intellectual approaches and techniques for making connections?[46]

---

[46]Sentman and Hope, 1994, "Disciplines in Combination," p. 6.

One illustration of how curriculum design is thought through is provided by the standards for joint accreditation by the National Association of Schools of Music and the Accreditation Board for Engineering and Technology of baccalaureate degree programs combining studies of music and electrical engineering. Addressing admission, faculty, facilities and equipment, library facilities, curricular structure, specific course requirements by category, and essential competencies, experiences, and opportunities, the standards spell out what would and would not be acceptable for joint accreditation, and they provide guidelines for the operation of "combination degree programs" in music and electrical engineering.[47] That effort illustrates implementation that responds to fundamental concerns raised by Sentman and Hope: "Connect the work of the arts unit to professional work in other disciplines. Teach by example the interconnections of the arts professions with other intellectual and professional activities . . . ."[48]

Of concern to the committee is the narrowness or shallowness of some of the new offerings, which seem little more than venues for learning and applying skills in certain software packages (e.g., PhotoShop), and the observation that in some contexts, the rise of new activities involving IT seems to dominate preexisting programs (with the acquiescence of those in charge of such programs), with questionable results for curricula and degree programs. While initial interest may begin with superficial use of IT, there is a need to think through, in the long run, the emphasis of academic activities on intellectual content relative to process: Generating in-depth knowledge, skills, and competence requires intellectual content, not just method.

Truly good curricula that support ITCP work are difficult to design, because they must reflect a sufficiently broad scope of learning and inquiry while also incorporating the core of some body of knowledge. Graduates are best served by gaining both exposure to ITCP work and grounding in some discipline.[49] The practical realities of the employment marketplace—which tends to reward specialization and degree name recognition—suggest that jettisoning any attempt at specialization is perilous for the student.

The time is ripe for academic experimentation with ITCP, given broad trends among disciplines. It is a time when disciplines are fragmenting or have become so porous as to lose their former shape and identity. First formulated in the early 19th century in Hegel's

---

[47]See *Handbook of the National Association of Schools of Music, 1995*, National Association of Schools of Music, Reston, Va. There is apparently only one program, at the University of Miami, that conforms to the guidelines at this time.

[48]Sentman and Hope, 1994, "Disciplines in Combination," p. 7.

[49]Under ideal conditions, students would master two bodies of knowledge: computer science and a field within the arts or design. However, such a course of study is probably time (and cost) prohibitive for most students and universities.

Berlin, the now classic university fields and professions are undergoing a complex process of reconceptualization and restructuring. Just as the opposition between brain and body is being rethought, so, too, is the binary model separating the arts from the sciences, culture from nature, and biological from engineered systems.[50]

---

[50]For the last quarter of a century, philosophers, cognitive scientists, and cultural historians have been delving into the problem of human knowing and consciousness. George Lakoff and Mark Johnson summed up this turn toward neuro-inquiry as "philosophy in the flesh" (George Lakoff and Mark Johnson, 1999, *Philosophy in the Flesh: The Embodied Mind and Its Challenge to Western Thought*, Basic Books, New York). In light of recent developments in AI, robotics, biotechnology, IT, and the exponential expansion of the Internet, can one still claim, however, that "our reality is shaped by the patterns of our bodily movement, the contours of our spatial and temporal orientation, and the forms of our interaction with objects" (p. xix)?

# 7 | Institutional Issues and Public Policy

M
any of the opportunities and challenges associated with work in information technology and creative practices (ITCP) are presented in the preceding chapters. Chapters 3 and 4 focus on substance—the subject areas that merit attention—whereas Chapters 2, 5, and 6 focus on the processes and places that directly support ITCP work. There are also global factors that can indirectly promote or deter ITCP work that extend beyond the intellectual agenda and specific mechanisms and support structures previously described. Attention is focused here on the four relevant global issues on which committee members have informed commentary to offer: digital copyright, digital archiving and preservation, validation and recognition structures, and the geography of ITCP.

In these four areas, ITCP work could benefit from some type of concerted action—by government agencies; legislatures; courts; industries; colleges and universities; professional, scholarly, and trade associations; and other interest groups. First, the ongoing copyright debates on the use and re-use of digital information have important immediate and future consequences for the conduct of ITCP work. Second, the archiving and preservation of digital content for the benefit of future generations—to support both future enjoyment and to serve as a baseline for future ITCP work—requires action now before many digital works become lost. Third, new recognition and validation structures may be needed to evaluate and reward ITCP work that is notably different from mainstream information technology (IT) or arts and design work. And fourth, regional development policies and practices can encourage or discourage the evolution of environments conducive to ITCP work. The issue of funding from governments and private philanthropy is discussed in Chapter 8.[1]

• • • • • • • • • • • • • • • • • • • • • • • • • • • • • • • • • • • • • • • • • • • •

# DIGITAL COPYRIGHT

ITCP work builds on the cumulative record of social and cultural discourse, scholarship, and scientific debate and discovery. Simple and direct access to this record can greatly facilitate the production of creative work. Governments recognize the importance of this dependence in cultural, artistic, technical, and/or scholarly accomplishment. Intellectual property laws are fashioned with a careful balance of interests in mind—a balance that includes the interests of those who produce creative and intellectual work, those who wish access to it, and those who mediate between the producers and the users.

However, the advent of digital content and networks has upset this balance, which was crafted substantially for a world of physical artifacts, not electronic bits (see Box 7.1). Some suggest that the balance is skewed, and, in particular, that the balance has shifted in favor of commercial interests. Whether this shift has indeed taken hold can be debated,[2] but no one can deny that the interests of commercial content producers and distributors are well represented and highly visible in the public debate. Recently, however, other viewpoints are gaining prominence, through the work of scholars such as Lawrence Lessig.[3]

Capability in the creative use of IT needs to be understood as growing from the fertile soil of a common culture containing past creativity, ever more encoded in digital artifacts, and the nurturing of social processes, ever more mediated through IT networks, that make deep knowledge of this past creativity possible. The real choices available to individual ITCP practitioners in defining how their work circulates socially constitute an effective dimension of their creative freedom.[4] However, the mediation by networks raises new questions

---

[1] There are, of course, other public policy issues that have varying impact on ITCP work, such as the concentration of the mass-media industries and the role and impact of international organizations (some of which is discussed in Chapter 8 under the rubric of funding). The committee focused its attention on the set of public policy issues that it sees as most important to ITCP work and for which the committee is capable of providing specific commentary.

[2] See Computer Science and Telecommunications Board (CSTB), National Research Council (NRC), 2000, *The Digital Dilemma: Intellectual Property in the Information Age*, National Academy Press, Washington, D.C.

[3] See, for example, Lawrence Lessig, 2001, *The Future of Ideas*, Random House, New York.

[4] Michael Shapiro makes the important argument that while Americans usually do not think of their nation as having a cultural policy (since it prefers private over public funding models), in fact intellectual property regulation is precisely that—a politically

## BOX 7.1
### What Changed in the Balance of Copyright Protections?

• *Deterioration in the effectiveness of the natural inhibitors to copying.* Magazines, oil paintings, and most other physical objects involve non-trivial costs to copy, and in some instances, the costs can be prohibitive. Digital information largely overcomes these natural physical deterrents; copying bits is very inexpensive and typically only requires access to commonplace computer systems. The proliferation of digital information and networks makes it possible to produce and distribute many exact copies rapidly for nearly zero cost.

• *Deterioration in the effectiveness of the natural inhibitors to modification.* Similarly, modifying physically based objects often involves non-trivial costs (e.g., modifying a copy of a book), and the modifications may often be obvious once made. Modifications of digital information can usually be accomplished at very low cost and often are virtually undetectable.

• *Use of information technology to lock up digital information in ways that were previously impractical.* Information contained within physical objects (e.g., books) could not be reasonably controlled once distributed; by contrast, content-protection technologies could control access to some digital information indefinitely.

• *From display to publication?* The use of the Web to make work accessible can raise questions that did not arise previously. For example, consider a physical work of art that incorporates ordinary household items that are protected under copyright.[1] Copyright permission is not required to incorporate a soup can or soft drink container into an artwork. However, developing the comparable work for access on the Web may constitute publication and may therefore necessitate permission from the copyright holder (which also may involve the payment of a fee) before such work can be made publicly available.

• *Digital archiving and preservation involves copyright law in an intimate way.* As described in the section "Digital Archiving and Preservation," archiving and preserving digital works often involves making a copy of the original (and possibly modifying it) and providing access to the copied work. Archiving of physical objects such as sculpture or books rarely involved such concerns.

• *Technology facilitates reuse of information, but the legal and economic infrastructure does not.* Networks and digital information can enable the development of extraordinary information technology (IT) and creative practices (ITCP) work that draws on many other works from around the world. IT facilitates integrative, improvisational approaches. However, the infrastructure to secure the legal rights to use other works is a long way from supporting many such projects and is a deterrent to the creative process.

• *Increasing use of contract law (licenses) instead of some provisions of copyright law.* One provision of particular importance in the conduct of ITCP work, especially work that incorporates small portions of other works, is the fair use doctrine.[2] If licensing increasingly replaces fair use, ITCP work can be inhibited because of difficulties in securing permissions, as discussed above. Cost is also an issue. The fair use doctrine was included in the copyright law as a part of the deal in copyright protection in support of the public interest—that is, protection is granted, but the copyright holder agrees to limited free access. Insofar as the increasing use of licenses implies increased cost, ITCP work is deterred.

---

NOTE: The content of this box is derived from (1) two commissioned papers of the Art, Technology, and Intellectual Property Project of the American Assembly: "Looking Backward and Forward" by Michael Shapiro, 2001, and "Public Policy at the Intersection of the Arts" by Margaret Wyszomirski, 2001, available online at <http://www.americanassembly.org/ac/atip_p_cp.htm>; (2) the final report of the Art, Technology, and Intellectual Property Project of the American Assembly, 2002, available online at <http://www.americanassembly.org/ac/atip_na_fr.htm>; and (3) Computer Science and Telecommunications Board, National Research Council, 2000, *The Digital Dilemma: Intellectual Property in the Information Age*, National Academy Press, Washington, D.C.

[1] For some items, trademark law might be more applicable. Nevertheless, the basic point still applies.

[2] The fair use doctrine relates to a provision in U.S. copyright law that permits limited copying and use of works protected by copyright to be undertaken without the explicit permission of the copyright holder. The scope of "limited" is too complex to be addressed here; see Chapter 4 of *The Digital Dilemma*.

about rights and responsibilities that, depending on how law and norms evolve, could chill some explorations.[5] Thus, future stakes include access to and protection of content as well as the shaping of platforms for cultural expression.

The openness of some software communities to the free exchange of software-code components and the designation of digital art objects as common property for reuse have become a contentious aspect of digital culture in the past decade. Code sharing, especially for research and education, characterized the early years of computer science; restrictions on sharing grew with proprietary, commercial interests; and the "open source" (for open access to source code) movement arose as an attempt to reset the balance and foster alternatives to closed, commercial systems. Notwithstanding the partial origins of the open-source movement in the quest for highly reliable generic computing resources, like operating systems or file transfer services,[6] there are similar incentives for individual creators to contribute to a common art or design platform.[7] Such a platform can permit a rising tide of community performance in particular domains, such as computer animation or interactive art, while still preserving the ability of individuals to create their own works. Similarly, an open-source approach may enable development of tools specialized for art or design uses that would not themselves justify commercial development. Thus, as the case of the Max programming language[8] shows on a small scale, ecologically diverse mixtures of proprietary and free components have appeared with growing frequency in large, industrial-strength applications offered by vendors such as AliasWavefront, SoftImage, and AutoDesk. This phenomenon, and the interest in ITCP communities in open source, illustrate how creativity is a process combining cooperation and competition. More thoughtful approaches are needed than

---

shaped framework for cultural development. In this framework, citizens' knowledge of their creative rights and responsibilities becomes a core capability within an information society. By analogy to the rise of environmental consciousness, public policy defining such capabilities should no more be delegated to private media corporation special interests than should the setting of environmental policy be dominated by either the Sierra Club or General Motors. See Michael Shapiro, 2001, "Copyright as Cultural Policy," Center for Arts and Culture, Washington, D.C.

[5]For a concrete example, consider the robotic plants of Ken Goldberg (described in Chapter 6), or any other participatory work that is enabled by the Internet. Who retains the copyright for each individual contribution? If the answer is "each individual," then use of the work and the collective individual contributions could require copyright approval (or a licensing agreement) from each participant.

[6]For further information, see <http://www.gnu.org/>.

[7]Emblematic was the organizing of the "International Conference on Collaboration and Ownership in the Digital Economy" (known as CODE) held in Cambridge, England, in April 2001. See <http://www.cl.cam.ac.uk/CODE/>.

[8]See the discussion in Chapter 3.

simply to discourage or criminalize the copying of digital content.[9] And (longitudinal) study is needed to understand what, in fact, is the impact of open source on creativity.

Creators of all types, from economically marginal or folk producers to professional commercial artists, need to become aware of the range of licensing options and permissions already available to them within current legal frameworks. One of these is, of course, to choose to place works in the public domain to facilitate reuse, or to pool them as part of intellectual property conservancies.[10] There are initiatives to assess the options available for balancing individual creative incentives with the legitimate needs of citizens to share and cooperate in the building of a common digital culture.[11] For example, the American Assembly's Art, Technology and Intellectual Property (ATIP) project is a national effort by a non-partisan, public policy institution to address the impacts, challenges, and opportunities resulting from technological advances that are confronting the arts.[12] Case Western Reserve University has established the new Center for Law, Technology, and the Arts.[13] Continuing efforts are needed, particularly those that can offer new and bold ideas that are commensurate with the effects on copyright law and practice attributable to the shift from analog to digital information. The development of specific proposals that could be implemented deserves priority over general philosophical and conceptual discussions of the impact of copyright law on the conduct of ITCP work.

The brief discussion above on this very complex and important topic can only begin to outline the many facets of digital copyright and especially those facets that have the most impact on ITCP, which embrace the use, control, distribution, ownership, and creation of ITCP work. A more complete examination would extend to other aspects of intellectual property rights—patents and trademarks. Thus, the purpose of this section is not to provide a comprehensive analysis

---

> Copyright issues greatly influence the nature of ITCP work, and such issues deserve further examination and action.

---

[9]*The Digital Dilemma* (CSTB, NRC, 2000) documents the debates about how to frame the commonly agreed upon importance of public education about intellectual property. *The Future of Ideas* (Lessig, 2001) leads the charge in articulating the counter-productive effects on the public welfare from imposing stringent intellectual property protection by making copyright valid for longer periods.

[10]Based on a presentation by Rick Prelinger, Prelinger Archives, at the committee's January 2001 meeting at Stanford University.

[11]CreativeCommons.org has launched a set of tools and templates for guiding creators in the "specialized" licensing of their work—suggesting intermediate solutions between the monopoly status of copyright and the uncertain and sometimes inappropriate status of general public licenses and other comparable mechanisms. "The project will make use of one of the techniques employed by digital rights management systems—specifically, the concept of a rights expression language—but with the primary goal of releasing (rather than reserving and enforcing) intellectual property rights" (private communication, Molly Shaffer Van Houweling, University of Michigan Law School, April 15, 2002).

[12]See <http://www.americanassembly.org/ac>.

[13]See <http://lawwww.cwru.edu/academic/lta/introduction.htm>.

of the copyright issues relevant to ITCP work, but to convey to the reader the message that copyright issues greatly influence the nature of ITCP work and that such issues deserve further examination and action. For detailed discussions, the reader is advised to review the works referred to in this section.

• • • • • • • • • • • • • • • • • • • • • • • • • • • • • • • • • • • •

# DIGITAL ARCHIVING AND PRESERVATION

In theory, digital information is perfectly reproducible and therefore potentially eternal—but practically speaking, this is far from the case. Instead, art and memories that are committed to digital media put the cultural heritage of the nation and the world at risk. Collection and exhibition must necessarily evolve into something that looks and feels like stewardship, not mere storage. As concluded in previous Computer Science and Telecommunications Board reports, new technical infrastructures are required to archive and preserve digital content,[14] or when feasible, to incorporate provisions for long-term preservation into ITCP work as an integral component of initial design. The challenge of digital archiving and preservation is formidable and cannot be addressed fully in this report. There are no easy answers. However, some of the most pressing problems can be outlined here, underscored by the committee's sense that urgent action is needed.[15]

Physical artifacts, although they require environmental controls and secure storage, can be preserved more easily than digital information because their formats do not become obsolete over time—a painting or sculpture remains "readable" for centuries into the future. Digital content, on the other hand, can become difficult to read in a matter of a decade (or even less) as formats and systems for digital content evolve.[16]

---

[14]For detailed treatments on digital archiving in general, see Computer Science and Telecommunications Board, National Research Council, 2001, *LC21: A Digital Strategy for the Library of Congress,* National Academy Press, Washington, D.C.; CSTB, NRC, 2000, *The Digital Dilemma;* and the forthcoming report of the Committee on Digital Archiving and the National Archives and Records Administration (for further information, see <http://www.cstb.org/project_nara.html>).

[15]The next three paragraphs are based largely on Howard Besser's "Longevity of Electronic Art," February 2001, available online at <http://www.gseis.ucla.edu/~howard/Papers/elect-art-longevity.html>.

[16]Try to find the hardware to read a Multimate (word processor) file on a 5.25-inch floppy disk saved using the CP/M operating system. Howard Besser characterizes this situation as "the viewing problem"—that is, "the default for electronic objects is to become inaccessible unless someone takes an immediate pro-active role to save them." See Besser, 2001, "Longevity of Electronic Art."

Files can be moved from one physical medium to another (called refreshing) in an effort to avoid ending up with files that are increasingly difficult to read because of technological obsolescence or physical decay of the medium. Refreshing does help, but it does not address the problem of evolving file formats. Refreshing files in obsolete formats (e.g., Wordstar word processing files) still leaves the problem of being able to read the obsolete format. The problem of obsolete formats becomes more difficult with multimedia information, as the different modes (sound, text, images, and so on) are represented using different formats.

As a file of bits, the physical rendition of an electronic work is not fixed. ITCP works may be viewed on a personal computer with an Internet Explorer or Netscape browser today, but might be viewed using very different hardware in the future (or even the present)—with each configuration of hardware and software offering a different user experience, though based on the "same work." For example, faster hardware and networks can cause works to operate at a more rapid pace than intended by their creators and, ultimately, as hardware and networks become orders of magnitude faster, the user's experience could change markedly. Curators and conservators will need to identify those characteristics of the enabling information technology that are integral to a work.

Even the definition of a work can be problematic. A work may link to other works (that may belong to the same creator or not), and those other works may be live works themselves, that is, changing over time in response to users' actions, be they descriptive or structural in effect. Capturing everything necessary to preserve works that link to other works can be quite difficult—in a technological and institutional sense.[17] Determining whether a work is authentic also becomes more challenging: The changing of a few bits within a work (whether by accident or design) can alter a work in a substantial way and can be extremely difficult to detect.[18]

Some ITCP works are created to interact with people—such works are alive in the sense that a sculpture or painting is not—and therefore these works can change over time by intent, presenting the problem of how to archive and preserve multiple versions of ostensibly the same work. In this way, ITCP work can be more challenging to archive and preserve than some other forms of digital content that are, for instance, fixed (e.g., electronic versions of printed scholarly journals).

---

[17]There are also legal questions. Although one may link to the work of another, copying another work as an integral step in the preservation of a given work (which presumably will be made accessible to various people in the future) may well violate copyright law (for works under copyright protection), at least in the United States. See the section "Digital Copyright" above in this chapter for further discussion.

[18]For example, consider a computer game in which several key parameters are changed. In terms of how the game looks and plays, all may seem as usual, except to an enthusiast who has played the game many times and can detect that "something does not feel right" or that it seems a lot harder to achieve some particular goal than previously.

It is not uncommon for ITCP-based work to end its life with the close of the original project and exhibition, much before the challenges of obsolete hardware, software, and file formats become relevant. Therefore, archiving and preservation of ITCP work depend at least in part on the status of new-media art within curatorial institutions. For example, digital works that enter a permanent collection presumably will be properly maintained and migrated to new digital formats as necessary to ensure that the work remains accessible using reasonably available information technology. Jon Ippolito, associate curator of media art at the Solomon R. Guggenheim Museum, said, "The objective is both to demonstrate our conviction that these forms of cultural expression deserve to be safeguarded for the future and also to demonstrate a method for doing it," when commenting on the Guggenheim's acquisition of its first two works of Internet-based art for its permanent collection[19] in the beginning of 2002.[20] Another recent initiative is the project Archiving the Avant Garde, which intends to establish guidelines for museums, galleries, and artists for preserving art that incorporates information technology in significant ways.[21] Other notable archiving initiatives include the Conceptual and Intermedia Arts Online project (CIAO)[22] and the Variable Media Network.[23] The Daniel Langlois Foundation houses the Centre for Research and Documentation, which has created an online database of artifacts that documents the history, artworks, and practices associated with digital media arts and then makes the information available to researchers and the general public.[24]

Electronic literature provides a good example of the problems and possibilities. The practice of literature and literary criticism would be almost unthinkable without the rich resources of the codex archive, stretching back to early print books and beyond that to manuscript culture. Recognizing its crucial importance, major institutions have devoted significant resources to the preservation and accessibility of this archive, including great public and private libraries such as the British Museum Library, the Getty Research Library, and the Beinecke

---

[19]The works are net.flag by Mark Napier and Unfolding Object by John F. Simon, Jr., available online at <http://www.guggenheim.org/internetart>.

[20]Derived from Matthew Mirapaul, 2002, "Getting Tangible Dollars for an Intangible Creation," *New York Times*, February 18, p. B2.

[21]Participants in the Archiving the Avant Garde project include the Walker Art Center, Cleveland Performance Art Festival and Archive, Solomon R. Guggenheim Museum, Franklin Furnace Archive, Berkeley Art Museum and Pacific Film Archive, and Rhizome.org; see <http://www.bampfa.berkeley.edu/ciao/avant_garde.html>. Also see Scott Carlson, 2002, "Museums Seek Methods for Preserving Digital Art," *Chronicle of Higher Education*, May 28, available online at <http://chronicle.com/free/2002/05/2002052802t.htm>; and Kendra Mayfield, 2002, "How to Preserve Digital Art," *Wired*, July 23, available online at <http://www.wired.com/news/print/0,1294,53712,00.html>.

[22]See <http://www.bampfa.berkeley.edu/ciao/>.

[23]See <http://www.guggenheim.org/press_releases/newmedia_pr.html>.

[24]See <http://www.fondation-langlois.org/>.

Rare Rook and Manuscript Library at Yale University. But there is as yet no national program for the archiving and preservation of electronic literature and art. As indicated above, the Guggenheim has begun to explore the problem as it applies to electronic art, but there is as yet no consensus on what and how to archive these works. Solutions include a migration strategy, which would require periodic updating and translation of works to ensure that they can be implemented on contemporary software and hardware. Emulation focuses on the development of software that can read and interpret files in older formats. Another possibility is archiving of the machines and software on which the works run, an approach that would require continuing maintenance that becomes increasingly problematic as the machines age, as well as increasing amounts of storage room and accessibility. Yet another possibility is to identify an extensible coding language, such as the extensible markup language (XML), and publish standards that writers and artists would be encouraged to follow if they wanted to make their works as accessible as possible for archiving. In this case, concerns include securing the equivalent of electronic museums on the Web, where works could continue to be updated and made accessible to users as the coding languages continue to evolve. The Electronic Literature Organization[25] is exploring all these possibilities, but again no consensus has yet emerged that is likely to win support from the relevant stakeholders.

> **There is as yet no national program for the archiving and preservation of electronic literature and art.**

• • • • • • • • • • • • • • • • • • • • • • • • • • • • • • • • • • • • •

## VALIDATION AND RECOGNITION STRUCTURES

Validation and recognition structures are needed to encourage the production of good work in ITCP, to offer reliable indications of quality to users, the public, and funders, and in academia, to support decisions on hiring, tenure, and promotion.[26] The nature of the output (consider the cliché, "but is it art?") and the nature of the methodology (i.e., the actions taken to integrate IT into art or art into computer science) may raise questions among reviewers.

A successful artist must satisfy two audiences: a public audience and the elite circle of critics, theorists, dealers, curators, and collec-

---

[25]The mission of the Electronic Literature Organization is to "facilitate and promote the writing, publishing, and reading of literature in electronic media." See <http://www.eliterature.org>.

[26]As one reviewer put it: "The new media—and simulations of the old—that computers can create allow immense novelty. But it would be surprising if the ideas on these fringes are less mediocre than most ideas elsewhere. In fact, one might expect the opposite: that it will be very difficult for most who are trying to express through the computer to find and invent genres that can carry real content expressed in new and important ways."

tors.[27] If an art piece gets into juried competitions because jurors like it, and then gets into galleries where many viewers see it and like it, then a verbal and written discourse is generated that renders a judgment. Critical acclaim is a strong motivator and criterion for success for much cultural production.[28] More prosaic criteria are also applicable: access measurements such as exhibition attendance or Web site hits or commercial success of artists. Designers engaged in ITCP represent a kind of hybrid of artists and information technologists—market appeal is a criterion of success, but so is positive feedback from the circle of elites. Although there may be a discrepancy between popular and elite appraisal at any time, either or both may become more positive well after a work has been made public, even after the creator has died.

Computer scientists (information technologists) who are employed by industry—or leverage their results by starting up a new business—may rely on market feedback. The late-1990s surge in entrepreneurial activity by computer science faculty suggests a swing toward market acceptance as a validating factor in that time period, although computer scientists continue to view academia as the locus of fundamental research and idea generation and industry (the dot-coms and other businesses) as the locus of the application of these ideas—the implementers. Market acceptance may generate professional recognition in retrospect, as is clear in the history of both computer science and the arts.

On the arts side, analysts and critics—such as art historians—have long contributed to deeper understanding than mere access to artistic artifacts provides. There appears to be a lag, however, in analytical appreciation for ITCP. One reason is the prominence of fast-moving popular culture, although the history of art shows that forms must be stable for at least a while before they yield profound expression. Another reason is the slow embrace in the arts-analytical community;[29] and a third reason is the undocumented nature of ITCP, which confounds scholarship. (As noted in the previous section, it may well be that humanistic scholarship in this arena will be stalled until strong, enduring alliances can be made with information scientists and com-

There appears to be a lag in analytical appreciation for ITCP.

---

[27]A reviewer explained this challenge, a kind of double standard, as follows: "In the popular imagination (and in the popular press) the work is no good unless it is accessible to the general public. But professional advancement is the result of approval among the elite circle of critics, theorists, curators, dealers, and collectors. Imagine if a condition for the Nobel Prize for physics was that, in addition to being groundbreaking science, Joe-on-the-street must understand and like it!"

[28]A practical challenge for all parties addressing IT and creative practice is maintenance of a critical attitude toward the technology. This challenge is exacerbated by the sheer difficulty and time demands of developing, maintaining, and updating technical skills, and of building reliably functioning technology. Artists' propensity for social commentary can shape attitudes toward technology (and technologists).

[29]This was remarked upon by several reviewers.

puter scientists, as these two groups become increasingly familiar with the new properties of emergent digital cultural forms and thus are able to help develop a scholarly apparatus appropriate to these forms.) The lack of historical perspective among practitioners and appraisers of ITCP—that is, the lack of an understanding of the durability of new ideas and their long-term impact, and also of an understanding of how and when ideas emerged and evolved—complicates validation and recognition and contributes to considerable wheel-reinventing.

Work in ITCP can sometimes be given acclaim as breakthroughs or as pioneering new forms of work. Being first to do something is often valued in technical contexts but may obscure the artistic merits. In this situation, the purpose of a work "often seems to lie outside the experience of the artwork as actually encountered. Instead the caption and illustration in the catalog, Web site, or newspaper seem to play a far greater role in the project's perceived success than the actual experience it engenders. The labels and photographs seem to push the artwork along every institutional step of the way: before, in grant applications; during, for publicity; and after, to help memorialize the work, readying it for its incorporation by curators, critics, and academics into a historical timeline."[30] Part of the problem lies in the nature of IT, which confounds early identification of innovations that will have a lasting impact: The computer (hardware plus software) is a very plastic medium, inviting the creator to push the edge of the current convention and side-step the current limits of the medium. This is an invitation to move fast, to leave the past behind, and to go explore a constantly new future. In this early period of experimentation, it is hard to tell what is really creative, what will end up being of lasting value, what is passing novelty, and so on; too much is in flux to sustain the usual critical feedback processes. This situation helps to explain why it seems to take a long time to integrate a new technology into making non-trivial art. Often in the past, it has been the children who have grown up in a genre who have produced the most profound expressions. The innovators have tended to be more awkward, even in the so-called classical disciplines, while the second generation has often gained the fluency to push the ideas the furthest. Most forms that we make things from—whether languages or materials—have more degrees of freedom than most of us can easily handle. This gives tremendous freedom and range that have to be balanced with considerable discernment, taste, and criticism, whether one is a painter, composer, or computer scientist.

In industry and in various design and art contexts, assessments may be less rigid as compared with those in academia, inasmuch as there are greater total numbers of jobs or other outlets for work. Even

---

[30]In the words of a reviewer of this report.

so, evaluations vary based on whether a new idea is seen as "just" an application or an incremental advance rather than as a significant innovation—the term "creative" is more likely to be used to describe the latter. In both academic and other contexts, there may also be an emphasis on whether an individual (or team) is the true originator of a new idea. Mass-market acceptance of new ideas and output/products (as opposed to acceptance by a community of peers) is not sufficient; it may be acknowledged within and outside academia, but it is no guarantor of high professional or peer regard,[31] although it may matter more for some subdisciplines (e.g., design, computer graphics) than others. Also, as explained by a member of the committee, "Popularization and elite curation, though seemingly opposed, actually work together to produce stronger results than either alone could."

The extant structures for computer scientists, artists, and designers are each based on the differing goals and motivations of the respective fields; the criteria used in validation and recognition structures relevant to ITCP also differ by subdiscipline within those arenas; and validation and recognition structures for ITCP work should reflect multiple sets of goals and motivations. Differences within a field in how people perform research or undertake specific practices are also difficult to appreciate across disciplinary boundaries. This confounds the appraisal of creative practices, or excellence. In computer science, how does one judge whether a particular breakthrough is creative or not?[32] Even among the arts (or the sciences), it is no easy matter to appraise creative practices across disciplines; one should not assume that a painter would automatically recognize when a jazz pianist is particularly creative, or that a network expert could easily discern a creative advance in computer graphics.

Review panels, whether associated with a conference, a professional society or group, or a publication, tend to be composed of evaluators with expertise in an established field, discipline, or profession. Since ITCP work often crosses established boundaries of expertise, panels can be unprepared to conduct proper peer review of that work, especially when it includes substantive technical and cultural, critical, artistic, or design components. Alternate mechanisms for the review of ITCP work are needed. For example, one possibility is to establish small-scale independent workshops for this purpose under the auspices of several prestigious organizations (to help establish legitimacy and recognition) in diverse domains. Mechanisms to support self-organization among those pursuing ITCP work would also be helpful.[33]

---

[31]In a famous example, astronomer Carl Sagan (creator of the television series *Cosmos* and the book and movie *Contact*) was probably the most popular science communicator of his day, but he was not elected to the National Academy of Sciences.

[32]An experiment by the committee, featuring a set of theoretical computer science examples, yielded mixed reactions (at best) even among computer scientist reviewers.

[33]See Chapter 5 for a discussion of possible mechanisms.

Maturation of

appraisal criteria

must develop

along with ITCP

work.

Ultimately, a set of distinctive goals and motivations may emerge for ITCP work, making it easier for leading practitioners to achieve the recognition that they deserve. Maturation of appraisal criteria must develop along with ITCP work, and a critical mass of critics sufficiently sophisticated to evaluate ITCP work is needed. An important factor in these developments will be the evolution of a language of discourse. Achieving great art in the future, regardless of the medium, implies finding enough stability in the language or media carriers of the ideas for both creators and appreciators (including critics and curators) to learn them. The evolving language of ITCP, of course, must be one that bridges multiple disciplines, and that bridging will have to acknowledge the different meanings attached by computer science and the arts to such words as "abstract" or "representation."

## PUBLICATION

Publication is important to both computer science and the arts and design (though probably more important to computer science), and this activity is evolving along with the role of technology as both input and output element. Publication vehicles vary in the extent to which they are perceived as indicators of validation and recognition. The gold standard for academia—and the criterion most easily understood by parties outside a given subdiscipline—is the so-called archival journal (often published by scholarly or professional societies) that involves considerable editorial selection plus prepublication review and revision, which function as a screening system for quality. But the long lead time for such publications poses problems for subdisciplines in which timeliness—quickly getting an idea into the field—matters. In experimental computer science, for example, the need for speed has led to an emphasis on conference publications for disseminating new ideas (journal articles are preferred for summarizing and consolidating a long-term body of work, and they are more amenable to longer write-ups). Conference proceedings offer greater speed of publication at the expense of broader audiences, stronger refereeing, and recognition associated with archival journals. Yet the prestige associated with presentations at major conferences actually makes some of them more selective than journals, especially in computer science.

Outside of academia, the range of relevant publications increases markedly. Publications can range from *Wired* and *RES* to *Artforum* and *Vogue*. Anthologies, surveys, and monographs cover the work of artists. Catalogs are created from museum exhibitions, film festivals, and gallery shows. Works that are performed or displayed publicly may be reviewed by critics, providing a public forum of validation and recognition.[34] Such reviews can appear in specialized professional or trade periodicals or in the popular press—from the *New York Times* to a small-town newspaper.

---

[34]In contrast to the peer review process that often provides anonymous feedback.

Broadening use of the Internet has spawned new online publications and induced changes in journals. Many print journals have now gone to electronic format, either to supplant or to complement print editions. These journals typically adhere to the same gate-keeping mechanisms, including peer review, that they developed for print. Some journals that publish exclusively in electronic form, for example *Postmodern Culture*,[35] also have gate-keeping mechanisms equivalent to those for print. Others occupy a position midway between self-publishing and reliance on peer review, having submissions reviewed by the editors but not by outside readers. As long as editorial policies are clear to users, these electronic journals, whether partially or fully reviewed, can perform valuable services. Similar observations can be made about "technical reports" in computer science, which are usually published by the researcher's organization and typically do not involve peer review. Web sites published through individual initiatives, lacking any validation mechanism, can have content that ranges widely from the authoritative to the misleading.

## CURATORIAL WEB SITES

For IT-informed literary ventures and other forms of creative expression, Web-specific practices are needed for carrying out some of the reviewing functions that have developed over three or four centuries for artistic production. This is especially important given the surge in volume of material associated with the enabling of creative efforts by amateurs. One way this concept is being implemented is through curated Web sites that screen large amounts of material, evaluate the products, and then feature only those deemed to be the best. These sites combine traditional reviewing functions with the collection functions served by galleries and museums, where visitors can be assured that they will be able to see several pieces gathered together in a single physical location. As curated Web sites gradually gain credibility, they serve as important display sites for audiences interested in electronic art but unsure where to find it. Existing sites of this type include those curated by Marjorie Luesebrink, Jennifer Ley, and Carolyn Guertin that assemble the best of Web-specific electronic literature: "Progressive Dinner Party" and "Jumpin at the Diner."[36] "Best of the Web: Museums and the Web," a juried competition coordinated by Maria Economou, identifies museum winners in the categories of online exhibition, museum's professional site, educational use, and research site.[37] Even individually curated sites can provide valuable recognition; for example, the Digital Libarian site[38] produced

---

[35]See <http://www.iath.virginia.edu/pmc/contents.all.html>.

[36]For further information, see <http://califia.hispeed.com/RM/predinner.htm> and <http://califia.hispeed.com/Jumpin/>.

[37]See <http://www.archimuse.com/mw99/best>.

[38]See <http://www.digital-librarian.com>.

by Margaret Vail Anderson, a professional librarian, recommends a wide range of library resources.

Other organizations are using the traditional commission and curatorial model for determining which individuals will receive electronic archive and display space. Although this strategy may limit the number of digital pieces available for presentation on a site, it does, through the commission function, provide the financial assistance needed to construct the work. Organizations using this strategy include Turbulence,[39] a Web site sponsored by New Radio and Performing Arts Inc., and The Alternative Museum (TAM).[40] Using a peer-review process, Turbulence selects up to 20 Internet art projects per year to commission and display on its Web site. In return for providing the commission and display space, Turbulence retains exclusive rights to display of the work for 3 years. Similarly, through its Web site, TAM sponsors the Digital Media Commissions program. Each year, with the help of a select committee of professionals from arts and technology fields, TAM chooses artists for whom it provides technical and financial support for the production of Internet art. Once completed, the projects are displayed together as an online exhibition.

Community-based, grass-roots approaches to filtering for content, quality, and relevance are made much more feasible by the Internet. Such approaches are discussed in Chapter 5.

## AWARDS AND PRIZES

The value of awards and prizes as a means to recognize outstanding achievements and stimulate further creativity in research is well known, but the numbers of prizes available are unevenly distributed across disciplines. Awards and prizes can be based on recognition—accorded for past accomplishment—or can be offered to motivate the creation of new work (e.g., an award based on the best entry at an exhibition). Some fields, such as chemistry and literature, have a number of prizes awarded by professional organizations,[41] whereas other areas—including computer science—have relatively few. This situation places an extra premium on the ability to secure funding (e.g., grants, employment in a suitable job), a process that, under contemporary conditions, can inhibit risk taking. Especially needed in the present climate are awards specifically recognizing creativity in both artistic and technical areas. Noteworthy in this regard are the Electronic Literature Organization's awards for the best work of electronic fiction and electronic poetry, which in 2001 provided substantial prizes of $10,000 each. Also exemplary is the Lyman Award, recently funded by the Rockefeller Foundation and administered by the National Center

---

[39]See <http://www.turbulence.org/>.

[40]See <http://www.alternativemuseum.org/>.

[41]In addition, these established fields (and some others) derive further legitimacy through awards such as the Nobel Prize, the National Medal of Arts, the National Medal of Science, and the Pulitzer Prize.

for the Humanities, recognizing the creative use of information technologies in the humanities, with a prize amount of $25,000. In 2001, the Prix Ars Electronica competition celebrated its 15th anniversary; it is a highlight of the Ars Electronica Festival program. From 1987 to 2001, 169 of these prizes with a total value of approximately $1.2 million were awarded.[42] However, such opportunities for recognition in the digital arts remain rare. See Chapter 8 for a discussion of awards and prizes as a means of financial support for ITCP practitioners.

## THE GEOGRAPHY OF INFORMATION TECHNOLOGY AND CREATIVE PRACTICES

Particular geographic configurations can support creative work in profound ways, even with information networks linking remote parties. There are hot spots of scientific, technological, scholarly, and artistic creativity, as exemplified by ancient Athens and Renaissance Florence, just as there are large areas where very little activity of this sort takes place. The explanations for this uneven spatial and temporal distribution are far from simple. As Harry Lime provocatively (and no doubt unfairly) observed in *The Third Man*, "In Italy, for thirty years under the Borgias, they had warfare, terror, murder and bloodshed. But they produced Michelangelo, Leonardo da Vinci, and the Renaissance. In Switzerland they had brotherly love. They had five hundred years of democracy and peace. And what did that produce? The cuckoo clock."[43]

However, Peter Hall has convincingly made a case, in *Cities in Civilization*,[44] that creativity flourishes mostly in urban settings. Cities allow the specialized division of labor that is necessary for intellectual exploration and innovation. They can support specialized facilities, such as libraries, that creative work frequently demands. And they provide the conditions for effective collaboration among specialists.[45]

---

[42]See <http://www.aec.at/> for further details. The Prix Ars Electronica competition attracted controversy in 2000 when the writer Neal Stephenson won in the Internet category, even though he did not enter the competition. See Matthew Mirapaul, 2001, "And the Best Internet Art Is . . . Virtually Anything," *New York Times*, September 3, p. B2.

[43]From the 1949 movie produced by Carol Reed.

[44]Sir Peter Hall, 1998, *Cities in Civilization*, Pantheon Books, New York. Richard Caves has also examined arts clusters in specific cities. See Richard E. Caves, 2000, *Creative Industries: Contracts Between Art and Commerce*, Harvard University Press, Cambridge, Mass.

[45]State arts agencies track grants awarded to rural artists and organizations, which are an order of magnitude less (in dollars) than those going to urban areas. See State Arts Agencies, 2002, "State Arts Agency Funding and Grant Making," National Assembly of State Arts Agencies, February.

Information technology has changed the established geography of creativity in two ways. First, it has produced hot spots, such as Silicon Valley, that are clearly fueled by IT itself. Second, it has supported specialized division of labor on a global scale; made libraries and other facilities available online rather than at specific, privileged locations; and enabled geographically distributed remote collaboration. These geographical factors have implications for future progress in ITCP.

## INFORMATION TECHNOLOGY HOT SPOTS

The emergence of IT hot spots, in areas such as California's Santa Clara Valley, Boston's Route 128 corridor, and the Austin (Texas) vicinity, is a special case of the general phenomenon of industrial clustering. There are many examples of this. The automobile industry is clustered intensively (though not, of course, exclusively) in Detroit; the film and music industries in Southern California; the publishing industry, arts and design communities, and museums in New York City; the biotechnology industry in Cambridge, Massachusetts, and so on. However, it is worth noting that growth occurs both within and beyond the clusters.

Once a cluster is firmly established, it is not difficult to see how it creates a regional advantage. Silicon Valley has become a magnet for IT talent from all over the world. It provides specialized facilities and opportunities that are not available elsewhere. The concentration of talent and activity within a relatively small area produces valuable knowledge-spillover effects. And the talent and resource pool available in the area makes it relatively easy to put together cross-disciplinary collaborations to innovate. The classic conditions for creativity are in place.[46]

The formation of such clusters is a more mysterious process, and it often seems highly dependent on unique circumstances at particular locations—with one thing building on another.[47] In Silicon Valley, the story began with a tradition of technological innovation in the early 20th century and developed further with the creation of a high-tech industrial base around Stanford University in the 1950s—student networks and fluid relations between scientists and instrument makers were key factors in the area's vitality.[48] It continued with the growth of

---

[46]For further discussion, see AnnaLee Saxenian, 1994, *Regional Advantage: Culture and Competition in Silicon Valley and Route 128*, Harvard University Press, Cambridge, Mass.

[47]See Manuel Castells and Peter Hall, 1994, *Technopoles of the World: The Making of Twenty-First-Century Industrial Complexes*, Routledge, London, for stories of Silicon Valley and other such complexes.

[48]Presentation by Timothy Lenoir to the committee at its January 2001 meeting at Stanford University. Also see Timothy Lenoir, 1997, "Instrument Makers and Discipline Builders: The Case of Nuclear Magnetic Resonance," *Instituting Science: The Cultural Production of Scientific Disciplines*, Stanford University Press, Stanford, Calif.

defense electronics during the 1960s and the emergence of semiconductor manufacturing and the microprocessor era in the 1970s. The personal computer industry built on this foundation and established a culture of innovative consumer electronics development. All that electronic hardware created a demand for software, and the hard-core technologists added yet another critical cultural layer. Finally, venture capitalists, lawyers, journalists, and marketers were attracted by the booming start-up activity, and in turn greatly facilitated that activity. In Silicon Valley's boom years, this was a magical—and difficult to replicate—mix.

The possibility of replicating Silicon Valley's vitality, creativity, and economic success is obviously of great interest to policy makers who would like to produce similar success elsewhere.[49] It is easy to identify some of the necessary elements in the magic mix, but considerably more difficult to combine enough of these ingredients in such a way as to produce Silicon Valley clones at other locations. Attempts have been made, with varying degrees of success, in Northern Virginia (United States), Sophia-Antipolis (France), Tsukuba (Japan), Hsinchu (Taiwan), and many other locations. Smaller-scale attempts have also been made, and some of these have focused on ITCP, which is less capital-intensive than some forms of IT production and engages a broader skill mix. An example is the western Massachusetts area and its Massachusetts Museum of Contemporary Art (MASSMoCA) in a former factory complex, which has attempted to attract digital media firms to co-locate at the intersection of IT and the arts.[50]

A more subtle question is whether Silicon Valley itself can sustain its hot-spot status as conditions evolve and change over time. For example, the development of the World Wide Web and digitally distributed news, entertainment, education, and other content produced a need to engage graphic artists, designers, writers, and others with IT. These sorts of specialists frequently did not find the engineering culture and suburban ambiance of Silicon Valley particularly congenial; many of them preferred a more urban setting and a different set of cultural connections, and so they tended to cluster in locations such as the SoMa area of San Francisco, the SoHo area of New York, and the

---

[49]See David Rosenberg, 2002, *Cloning Silicon Valley: The Next Generation of High-Tech Hotspots*, Reuters, London.

[50]When MASSMoCA opened in May 1999, its mission was not only to function as a laboratory for contemporary visual and performing arts, but also to help drive the revitalization of the economically depressed town of North Adams, Massachusetts. The town had experienced a dramatic downturn in its fortunes as the national economy moved away from a manufacturing base toward information and technology services. See <http://www.massmoca.org/mission/index.html>. Also see Kristen Andersen, 2002, "Bangor Looks to Bay State Town as Model for Culture," *Bangor Daily News*, October 24. For a broader discussion, see "The Role of the Arts in Economic Development," Issue Brief, June 25, 2001, National Governors' Association Center for Best Practices, Washington, D.C.

greater Los Angeles region.[51] As if to respond to this question, a local philanthropic and community development organization, Cultural Initiatives Silicon Valley, was formed in the late 1990s to promote the arts as a spur to creativity and community; it even used IT to create a crude cultural policy simulation game to link the arts and culture to the economic well-being of the area.[52]

It seems likely that in the 21st century a key challenge for the world's established IT hot spots will be to add a vibrant layer of arts and design to the magic mix. If they are unable to do so, their advantage is likely to fade as the cultural content of digital products and services grows in importance relative to the more purely technical content. If a company is making chips or operating systems, then it is an advantage to be located in a strong engineering cluster; but if a company is making digital movies, it may be more of an advantage to be located in a cluster of artists, actors, and musicians.

> A key challenge for the world's established IT hot spots will be to add a vibrant layer of arts and design to the magic mix.

## GEOGRAPHICALLY DISTRIBUTED CREATIVITY

Traditionally, the creative energy of cities has depended on face-to-face contact. So has that of university campuses. Even Silicon Valley—which is a low-density, automobile-oriented suburban area—clearly depends for its vitality on its dense concentration of specialists who can readily interact with one another. But there is a downside to relying exclusively on physical proximity to foster creative interaction. It limits the available talent pool, limits access to specialized facilities, and limits cultural diversity. There can be strong reasons for creative talent to be based far from urban areas.[53] So there is a strong case for taking advantage of digital technology to add remote participants and resources to the mix. These will include talent from developing nations, where IT has been leveraged not only for public access, but also for creating new kinds of work.

---

[51]Richard Florida refers to these specialists as members of the "creative class," who "seek an environment open to differences. Many highly creative people, regardless of ethnic background or sexual orientation, grew up feeling like outsiders, different in some way from most of their schoolmates. When they are sizing up a new company and community, acceptance of diversity and of gays in particular is a sign that reads 'non-standard people welcome here.' The creative class people I study use the word 'diversity' a lot, but not to press any political hot buttons. Diversity is simply something they value in all its manifestations." See Richard Florida, 2002, "The Rise of the Creative Class," *Washington Monthly*, May, available online at <http://www.washingtonmonthly.com/features/2001/0205.florida.html>.

[52]See <http://www.arts4sv.org/>. The simulator was eventually packaged with a monograph that provides context. See John Kreidler, Kate Cochran, and Brendan Rawson, 2002, *Great Cities: A Laboratory for Cultural Policy*, Americans for the Arts, available online at <http://store.yahoo.com/americans4thearts/greatcities.html>.

[53]For example, a study of Canadian artists suggests that "landscape appeal" inspires work and influences the decision by some artists to seek rural locations. See Trudi Bunting and Clare J.A. Mitchell, 2001, "Artists in Rural Locales: Market Access, Landscape Appeal and Economic Exigency," *Canadian Geographer* 45(2): 268-284.

Consider, for example, the production of the present report. Although the National Academies are based in Washington, D.C., the members of the team that wrote the report were scattered widely over North America and Europe; it would have been far too limiting to rely only on participants from the D.C. area. Some of the creative interaction necessary to generate the report was accomplished through travel to meetings at various locations, but a great deal of it was accomplished electronically—through e-mail and Web interactions. And this, of course, is now typical in group writing projects.

In the field of architectural design, major architects now compete for work globally—not just in their local regions. They put together teams of consultants and specialized subcontractors, as necessary for the particular job, from around the world. And they rely on digital telecommunications facilities and increasingly sophisticated software to manage the geographically distributed, cross-time-zone process. As a result of the growing emphasis on remote collaboration, computer-aided design systems are starting to look increasingly like specialized Web browsers.

Just as effective face-to-face collaboration depends on having the right sort of physical setting, effective remote collaboration depends on having the right IT tools. This reality has created intense interest in electronic design studios[54] and similar facilities, and in computer-supported cooperative work tools—which may support either local or geographically distributed teams. These tools do, of course, vary a great deal by field; online collaboration in graphic design requires support that differs greatly from that needed for geographically distributed musical ensemble performance.

## TECHNOLOGY-SUPPORTED NETWORKS OF CREATIVITY

The emerging pattern of creative production, in the IT era, seems to be one in which local clusters form intense nodes of activity within geographically distributed electronic networks. This model allows the formation of effective linkages between complementary but distant clusters. Thus, for example, the film industry remains clustered in Hollywood, but there is an active electronic link to London's Soho area—which has particularly strong postproduction capabilities. Silicon Valley has a strong electronic connection to the complementary capabilities of Bangalore in India and the founding of a West Coast campus of Carnegie Mellon University.[55] Universities are beginning to form strategic alliances, supported electronically, with overseas uni-

---

[54]See, for example, Jose Pinto Duarte, Joao Bento, and William J. Mitchell, 1999, *The Lisbon Charrette: Remote Collaborative Design*, IST Press, Libson.

[55]See <http://west.cmu.edu>.

versities that have complementary strengths; the current Cambridge University–Massachusetts Institute of Technology alliance is a particularly interesting example. And industrial and financial corporations are increasingly aware of the advantages of locating activity clusters in the different markets and cultures they serve, and electronically linking them.

In summary, creativity continues to have an identifiable geography, and the advantages that derive from unique, local subcultures continue to matter. But IT has overlaid onto this geography new possibilities for the aggregation of geographically distributed talent and resources into creative combinations. And, in particular, local creative clusters can now extend their potential by strategically forming electronic linkages to other clusters—maybe very distant ones—with complementary capabilities. One can begin to think of these clusters as the specialized professional neighborhoods of the global village.

# 8 Supporting Work in Information Technology and Creative Practices

**S**upport for information technology and creative practices (ITCP) comes from many sources and is difficult to measure. Several questions complicate a full understanding of ITCP funding:

- What are the boundaries for such work? (What is an "information technology and creative practice" expenditure? What is not?)
- Where—in market-based and non-market activities—does ITCP take place?
- How much of what is spent on ITCP work lies embedded within (and may be difficult to disentangle from) more conventional forms of information technology (IT), the arts, or design activities or programs?
- What commercial organizations support ITCP as part of the way they do business? What commercial organizations support ITCP in other ways and for other reasons?
- What governmental and other non-profit organizations support ITCP work? Where does one look for such support?

All but the last of these questions are the focus of previous chapters, which provide the context for this one. Chapters 2 and 5 sketch the rise of commercial ITCP—and implicitly the rise of commercial funding for ITCP—through new approaches to design (e.g., industrial design, architecture), targeted corporate engagement of artists (e.g., artist-in-residence programs), and new kinds of products and processes in industries that produce creative content (e.g., video games, animated film, music).[1] Because commercial activity spans only a

---

[1]One can observe (relative to the overall state of the economy) healthy and even growing industrial bases for these activities, but existing data on revenues or even employment in these areas do not allow easy inferences as to how much was spent on developing and applying ITCP. Detailed economic analysis was beyond the scope of this project.

portion of ITCP,[2] commercial resources are not sufficient to sustain ITCP, to make the most of its potential, or to broaden access to its benefits. Further, commercial activity is not evenly distributed. For example, the market for computer music is much smaller than that for computer graphics, which itself is skewed toward entertainment products. As in other arenas, non-market resources can often be invaluable in exploring areas where a market has yet to be or cannot be established. Also, there are non-market, public policy reasons for supporting ITCP activity, or the infrastructure for such activity, as discussed in Chapter 1.

This chapter focuses on non-commercial—government and philanthropic—funding for ITCP because (1) it is linked to the most exploratory and least mission-constrained activity;[3] (2) in the context of academic institutions, in particular, it is linked to education and human capacity building, which benefits activity across sectors; (3) it is most likely to sustain the non- and pre-institutionalized activities that have been significant in early ITCP and are associated with a significant component of the arts; and (4) it is associated with a broader set of public-interest objectives than commercial funding (which tends to be linked to production and distribution of a product). Inasmuch as commercial activities are synergistic with those in non-profit contexts, spending on ITCP in any one arena may be leveraged elsewhere.

Although government and philanthropic funding for ITCP has a broader scope than funding linked to creating and distributing commercial products, it comes with a range of conditions. Its effectiveness increases to the extent that funds-seekers can "see" ITCP through the strings on a given pool of funds and decreases to the extent that funds-seekers see those strings as constraints on their creativity.[4] Committee member attitudes ranged from seeing no substitute for resources they could use at their discretion to accepting pragmatically the strings that would link activities in ITCP to funders' interests as well as their own.[5] The funding challenge lies in ensuring that practitioners and funders have enough common interests to nurture a vigorous spec-

---

[2]There are important differences between "art" and "craft," and commercial funding generally applies to "craft."

[3]An analogy can be made to fundamental research for information technology: IT research and development overall is spread across commercial and educational (non-profit) organizations. Almost all of the commercial activity supports development of products, while the most exploratory work—fundamental research—is associated with government-funded activity in universities. See Computer Science and Telecommunications Board, National Research Council, 2000, *Making IT Better: Expanding Information Technology Research to Meet Society's Needs*, National Academy Press, Washington, D.C.

[4]Artists have their own vision and agenda, which often does not coincide with what someone else wants or needs to have made at a given point in time. These realities put artistic creativity at odds with conventional market forces. See Richard E. Caves, 2000, *Creative Industries: Contracts Between Art and Commerce*, Harvard University Press, Cambridge, Mass.

[5]Their positions varied with the degree to which they saw themselves as artists, and among the artists, the degree to which they favor a conception of art as self-expression, versus more collaborative or socially shaped conceptions of art.

trum of ITCP activities via their combination of creative effort and wherewithal. The committee hopes that this report will encourage more funders to understand the value of ITCP and, given their starting point, either become more open to funding relevant activity (i.e., generation of ITCP work; its display, performance, or preservation; corresponding education, training, or physical infrastructure) or more informed in allocating the resources they can provide.

• • • • • • • • • • • • • • • • • • • • • • • • • • • • • • • • • • • •

# FUNDING IN THE UNITED STATES

The committee believes that the United States lags other countries (see the section "Funding in the International Context") in financial support for ITCP. This is a judgment,[6] based on member familiarity with initiatives and programs in different countries (which are often seen as the leading venues for producing or displaying ITCP—see, for example, descriptions of ZKM, Ars Electronica, and so on in Chapter 5); the existence of larger and more sustained public support programs for the arts abroad (notably in developed nations), which are comparatively open to ITCP; and the observation that information technology research programs have provided limited and largely incidental support to date. There are no consistent data that support a precise analysis of relevant funding in the United States, let alone across countries.[7] Major foreign-based activities that focus on ITCP appear to be components of national or regional leadership strategies abroad (often seen as competitive responses to U.S. technical leadership).[8]

Against this backdrop, the United States can leverage early efforts worldwide—which have helped to demonstrate the ITCP potential and experimented with different approaches to nurturing ITCP—to foster new activities that can elicit and sustain ITCP. Those activities can draw from the substantial base of computer science research support in the United States, a differentiator of the U.S. potential in many respects, to support the hybrid character of ITCP through more stable, focused funding. This is likely to occur naturally through the evolution of funding patterns for both the advance of computer science and

> The committee believes that the United States lags other countries in financial support for ITCP.

---

[6]A comprehensive quantitative analysis of spending trends was beyond the scope of the committee, and available data are not necessarily comparable across nations (a common problem in international comparisons that is aggravated in this instance by the role of tax-policy-induced private spending relative to direct public spending).

[7]U.S. data, for example, tend to aggregate relevant activity together with other, more conventional kinds. This problem is typical of economic measurements when new activities arise, often at levels that are too small to measure using conventional survey mechanisms.

[8]One illustration is Europrix, representing "selection and promotion of Europe's best in multimedia" (see <http://www.europrix.org>); various national initiatives, such as the establishment of ZKM (see Chapter 5), also have this character.

the advance of the arts and design, but it can also occur through the express initiation of focused programs and initiatives, which the committee believes it is time to forge. The latter is more direct, but it may be difficult to obtain government support at a time when research programs seem to be reorienting to respond to homeland security, while private philanthropy is constrained by smaller endowments, a consequence of the decline in the stock market in the 2000-2002 period.

## SOURCES OF FUNDS

**Typical arts**

**grants are in the**

**low five figures;**

**typical computer**

**science grants**

**are in the**

**six figures.**

Absent detailed data specific to ITCP,[9] the funding potential for ITCP can be appreciated by examining the historic bases of funding for work in the arts, computer science, and other elements of IT R&D.[10] Arts funding is dwarfed by the funding for computer science and other IT-relevant research—as one committee member put it, there is "mysticism and longing on the part of artists when it comes to scientific funding." Typical arts grants are in the low five figures (using data sets that begin at the $10,000 level); typical computer science grants are in the six figures. In the aggregate, federal appropriations to cultural agencies and organizations are comparable in magnitude to federal support for computer science research, but only a small fraction of the former supports the equivalent of research—the generation of new expression. Arts funding often focuses on display, performance, education, facilities, and other dimensions of public access— and accordingly is most likely to go to organizations rather than artists. By contrast, IT research funding is more often awarded to individuals (principal investigators) or groups of individuals. Relative funding potential also reflects this apparent rationale: Quality-of-life concerns such as widened access to cultural artifacts seem to motivate arts funding, whereas funding for technical research is motivated by a larger set of economic, social, and governmental concerns that together have resulted in higher levels of funding.[11] Recent efforts to

---

[9]The relative paucity of data on the humanities—funding or otherwise—is a well-known phenomenon. See, for example, Robert M. Solow, 2002, "Let's Quantify the Humanities," *Chronicle of Higher Education*, April 19, p. B20.

[10]Inasmuch as ITCP may embrace other forms of science and engineering (e.g., biology, mechanical engineering), other categories of research funding may also be relevant.

[11]According to Heilbrun and Gray, "Most analysts who favor public subsidies for the arts place a very high value on the objective of improving access for all the people. Because it is rooted in the U.S. egalitarian ethic, that position also enjoys wide political support. . . . Survey evidence from Australia and Canada indicates that the general public does believe the arts produce external benefits and is willing to make substantial tax payments to support them." See James Heilbrun and Charles M. Gray, 2001, *The Economics of Art and Culture*, Cambridge University Press, Cambridge, U.K., p. 243 and p. 250. *Funding a Revolution: Government Support for Computing Research*, (Computer Science and Telecommunications Board, National Research Council, 1999, National Academy Press, Washington, D.C.) describes the rise of funding for computer science research and the co-evolution of government, industry, and academic interests and activities over the second half of the 20th century.

strengthen the linkage of the arts to economic benefits may motivate greater funding[12]—but doing so causes unease among some artists who worry about the implications of suggesting that art must be instrumental. Much as in basic research in the sciences, the idea of experimentation and research in the arts as contributing to human understanding is often slighted when criteria are reduced to quantifiable return on economic investment. Of course, market support for ITCP—through design and other product-related activities—does nurture creativity in the context of some organizational objectives.

State and local governments, given their different emphases, and greater focus on quality of life (extending to economic development as well as elements of culture) relative to the federal government, are significant funders of the arts,[13] spending an order of magnitude more than the National Endowment for the Arts (NEA).[14] But they do not (with a few exceptions) fund computer science (or other) research. State arts agency spending varies considerably, ranging in 2002 from a high of $4.26 per capita (Hawaii) to a low of $0.28 per capita (Texas); eight states spent at least $15 million annually and nine less than $1 million.[15] In 2000, "media arts" accounted for 3 percent of state spending; the largest shares went to music (18 percent) and theater (14 percent). About 0.03 percent of state arts agency grants went to individual artists in 2000 (48 percent of those dollars were for the visual arts). Individual activities supported included fellowships (which constituted about half of the activities), residencies, artwork creation (about one-eighth of the activities), apprenticeships, and performance.[16]

---

[12]See, for example, National Assembly of State Arts Agencies, 2002, "The Arts in Public Policy: An Advocacy Agenda," *The NASAA Advocate: Strategies for Building Arts Support* VI: 1. Available online at <http://www.nasaa-arts.org>.

[13]By the mid-1970s, each state had an arts council. For an overview of the state of knowledge about state-level activities, pointing to ongoing research, see J. Mark Schuster, 2002, "Sub-National Cultural Policy—Where the Action Is? Mapping State Cultural Policy in the United States," working paper of the Cultural Policy Center at the University of Chicago, January. Available online at <http://culturalpolicy.uchicago.edu/workingpapers/Schuster9.pdf>.

[14]See Schuster, 2002, "Sub-National Cultural Policy," p. 9. Note that the aggregate state legislature arts appropriations have exceeded federal arts appropriations since 1985. See National Endowment for the Arts and National Assembly of State Arts Agencies, 2002, "State Arts Agency Funding and Grant Making," National Assembly of State Arts Agencies, Washington, D.C., February.

[15]National Endowment for the Arts and National Assembly of State Arts Agencies, 2002, "State Arts Agency Funding and Grant Making."

[16]In 1999, "state legislatures appropriated about $400 million in funding for the arts, while local governments spent in excess of $800 million." See National Endowment for the Arts, Center for Arts and Culture, 2001, *America's Cultural Capital: Recommendations for Structuring the Federal Role*, Art, Culture, and National Agenda series, Center for the Arts and Culture, Washington, D.C. Available online at <http://www.culturalpolicy.org/pdf/acc.pdf>. The National Assembly of State Arts Agencies has released statistical information on state spending for the arts that shows that "after nearly a decade of robust growth, legislative appropriations for state arts agencies (SSAs) contracted slightly

Private philanthropy also favors spending on the arts over computer science research. These conditions make public arts support in the United States quite decentralized, while also drawing a sharp contrast with public support for fundamental computer science research, in which the federal government plays a dominant role. Nevertheless, the rise of ITCP presents the prospect of including more arts-like activity as research, more research-like activity as art, and new mixes of funders for new portfolios of activity. Insight into ITCP may lead traditional arts funders to support it more and also may expand the availability of resources for technical research.

## Federal Funding for the Arts—The National Endowments

**Public arts support in the United States is quite decentralized.**

Most of the federal government's art spending—which exceeded $1.5 billion in 2001—supports major national organizations, which in turn award funding to other organizations and artists and cover their own operating costs. Major national organizations include the NEA (and its organizational sister, the National Endowment for the Humanities), the Commission on Fine Arts, the Institute of Museum and Library Services, the National Gallery of Art, the Smithsonian Institution,[17] and the Kennedy Center for the Performing Arts.[18] Federal funding for the arts also comes from a range of programs distributed among a remarkable range of federal agencies.[19] Federal arts funding emphasizes public presentation (e.g., performances and displays) and education; accordingly, funding for major federal arts institutions (e.g., the Smithsonian) dominates federal funding.

---

in fiscal year 2002 as both the national economy and state budgets softened. In fiscal year 2002, appropriations dropped from $446.8 million to $419.7 million. This marks the first time in six years that aggregate appropriations fell. However, appropriation declines of $21 million in California and $5 million in New York account for half of nearly all of this decrease. When they are removed from total appropriations, the aggregate remains flat at zero percent change." As discussed in Chapter 6, state programs (e.g., in New York and California) have provided some support for computer science research, including research with links to the arts and activities that fall under the ITCP umbrella.

[17]In 2001, the Office of Management and Budget proposed transferring Smithsonian research funds to the National Science Foundation (NSF) as a means of consolidating science research. Under such an arrangement, Smithsonian staff would apply for NSF grants. Reports from the National Research Council and the National Academy of Public Administration opposed such a proposal. See <http://www.nytimes.com/2002/11/02/politics/02SMIT.html>.

[18]State and local funding adds about 50 percent more to the federal contribution; see National Endowment for the Arts, Center for the Arts and Culture, 2001, *America's Cultural Capital.*

[19]The National Endowment for the Arts Web site provides links (<http://www.arts.gov:591/federal-opportunities02/b-federal.html>) to arts funding programs at widely diverse federal agencies ranging from the Federal Emergency Management Agency to the Department of Agriculture, in addition to the agencies that one expects, such as the National Endowment for the Humanities.

The NEA stands out as the largest single public funder of the non-profit arts. Its FY 2002 appropriation from the U.S. Congress was slightly over $115 million, which represented an increase of more than $10 million from the year before.[20] The NEA divides its grants into several broad categories, including grants to organizations, partnership agreements, leadership initiatives, and fellowships[21] (which award grants to individual artists). Typically, grant amounts range from $5,000 to $100,000,[22] with some requiring "at least a 1-to-1 match in non-federal funds." Examples of funded projects within these categories include, among other things, dance and theatrical performances, exhibitions, workshops, festivals, apprenticeships, master classes, educational activities for children, and "innovative uses of technology that make the arts more widely available."[23]

Despite the NEA's important role in administering federal support for the non-profit arts and the significant impact that it has had on American cultural life since its creation in 1965, there are some limits on its usefulness with respect to promoting ITCP. For example, more than 40 percent of the NEA's 2001 budget went to state and regional arts agencies[24]—organizations that tend to fund primarily traditional genres of art (e.g., dance, theater, visual arts, and so on) and to emphasize display, education, and performance. Continuing resource demands of traditional activities reinforce the absolute constraint of a limited budget.

The same legislation that created the NEA also created the National Endowment for the Humanities (NEH).[25] The NEH budget over the last few years has been slightly higher than that of the NEA at a

---

[20]It had been slashed by more than 40 percent for FY 1996 in the wake of a great deal of political and cultural controversy. NEA's budget exceeded $150 million per year in the early 1990s, until it was cut radically for FY 1996, remaining fairly static for the next few years. For specific appropriations data, see <http://www.nea.gov/learn/Facts/ApprHist.pdf>.

[21]These are the NEA's only programs that still award funds to individuals: Literature Fellowships, American Jazz Masters Fellowships, and National Heritage Fellowships. Grants range from $10,000 to $20,000 and are not open to applications; rather, these awards are based on nominations from the arts community and the public. In the aggregate, these programs represent approximately $1 million of NEA's annual budget.

[22]A listing of NEA grant awards since 1995 can be found online at <http://arts.endow.gov/learn/Facts/Contents.html>.

[23]From <http://arts.endow.gov/learn/NEAGuide/GTO.html>. An example of the technology awards is the $62,000 grant awarded in 2001 to the Deaf West Theatre Company in California. The grant was awarded to support the design of a backstage communication system for deaf and hard-of-hearing technicians, to install a computerized control board for technical effects, and to develop lighting mechanisms that automatically focus on signing interpreters or actors. The project also hopes to train deaf technicians for jobs in the theater and to enhance the theater experience for deaf audiences. For more information, see <http://www.deafwest.org/>.

[24]National Endowment for the Arts, [undated], "NEA Fact Sheet: NEA at a Glance—2001." Available online at <http://www.nea.gov/learn/Facts/NEA.html>.

[25]The National Foundation on the Arts and the Humanities Act of 1965 (P.L. 89-209).

little over $120 million,[26] although the amount has remained fairly static. The NEH provides grants for humanities projects in four areas: preserving and providing access to cultural resources; education; research; and public programs. NEH grants typically go to cultural institutions such as museums, archives, libraries, colleges, universities, public television and radio stations, and (unlike those of the NEA) individual scholars who apply for funds. Thus, the NEH may support ITCP in the context of creative writing and literature, for example.

## Indirect Public Funding for the Arts

According to Michael Kammen, the United States set the precedent for making charitable donations to arts organizations (as well as other non-profits) tax deductible:[27]

> [In 1917, the United States] became the first nation to allow tax deductions for cultural gifts to museums and nonprofit cultural organizations. The pertinent legislation[[28]] has been altered several times since, sometimes in ways that seem inconsistent to the point of being bizarre, but the operative principle has been an immense boon to cultural institutions. Moreover, the principle has become increasingly attractive to European countries during the past decade or so.[29]

The amount of support generated for the arts through tax incentives is generally not included in statistics on government funding for the arts, for, as a previously cited paper points out, "foregone revenues" are "notoriously difficult to measure precisely."[30] Nevertheless, that paper reports that according to "the best estimates, indi-

---

[26]National Endowment for the Humanities, 2002, "Summary of Fiscal Year 2003 Budget Request." Available online at <http://www.neh.gov/whoweare/2003budget.html>.

[27]"[W]hat happens under U.S. tax law is that the donor's tax liability is reduced by an amount equal to the donation multiplied by the tax rate in that person's marginal tax bracket. The higher the individual's marginal tax rate, the greater the tax reduction per dollar given away, hence the less the cost of the gift to the donor.... The amount of tax saved by the individual is also the amount of revenue lost by the government on account of the charitable deduction. It is this lost revenue that constitutes the indirect support given by government to the nonprofit sector." See Heilbrun and Gray, 2001, *The Economics of Art and Culture*, p. 257.

[28]For example, the Revenue Act of 1917, the Estate Tax Law of 1921, the Gift Tax Act of 1932, and so on. Note that in recent years changes in estate taxes have been contemplated that could jeopardize the incentive for charitable giving of all kinds.

[29]Michael Kammen, 1996, "Culture and the State in America," *The Journal of American History* 83(3): 791-814.

[30]Bruce A. Seaman, 2002, "National Investment in the Arts," Art, Culture, and the National Agenda Issue Paper #6, Center for the Arts and Culture, Washington, D.C., p. 22. Available online at <http://www.culturalpolicy.org/pdf/investment.pdf>.

vidual donors, foundations, and corporations gave more than $10 billion to arts, cultural, and humanities organizations in 1999,"[31] approximately five times the level of direct federal support.[32] This is a highly decentralized mechanism, extending to employer matching programs[33] and private individual largesse. Organizations may benefit more than individuals, inasmuch as organizations may be better positioned to handle the administrative aspects associated with tax-exempt status—but some of those institutions, in turn, fund individual artists.

## Funding by Private Philanthropy

Funding for artistic endeavors is available from individuals, foundations large and small, and corporations—in that order[34] and all shaped by tax policy.[35] In addition to conventional arts support from corporations, there is at least some corporate support specifically for ITCP in academia. For example, the Media Lab at MIT receives about 80 percent of its funding from corporations, which provide support as members of a consortium that shares interests in any intellectual property arising from the Media Lab's activities. The Media Lab appears to be unique in its draw—and dependence—on corporate philanthropy to support ITCP;[36] other academic programs (and individuals) associated with the arts receive targeted support from corporations, which is

---

[31]Center for the Arts and Culture, 2001, *America's Cultural Capital*, p. 4.

[32]Another analysis, examining data through 1998, suggested that private giving provides about 40 percent of arts and culture organizations' revenue. See Loren Renz, 2002, "The Foundation Center's 2002 Arts Funding Update," Foundation Center, New York, N.Y. Available online at <http://www.fdncenter.org>.

[33]For example, Texas Instruments, through its foundation, encourages and matches tax-deductible gifts in support of the arts and culture: "The Texas Instruments Foundation Arts and Cultural Matching Gift Program was established in 1979 to encourage Texas Instruments employees, retirees, and directors to contribute to the arts. It provides an effective way of assisting you in contributing to the quality of community life. Dollars you contribute to qualified organizations you wish to support will be matched by the Foundation on a dollar-for-dollar basis. The Foundation will match, one for one, each eligible tax-deductible contribution of at least $50. The total maximum per individual donor that will be matched is $10,000 per calendar year (January 1 through December 31). . . . Our cultural organizations are a vital part of the quality of life in our communities. Such organizations depend on private support to flourish. We are pleased through this matching gift program to join you in helping to strengthen cultural life." See <http://www.tialumni.org/tiaa/2000%20TI-26453%20Arts%20%20Culture%20MGP%20form.pdf>.

[34] Renz, 2002, "The Foundation Center's 2002 Arts Funding Update."

[35]See Heilbrun and Gray, 2001, *The Economics of Art and Culture*.

[36]There is the question of whether the corporate support provided to the Media Lab is motivated primarily by intellectual property concerns, in which case the funding might be better construed as investments rather than as philanthropy, or is motivated by a larger rationale for support of academic research (which includes utilitarian motivations such as access to students and graduates), in which case philanthropy might be an appropriate characterization.

**The economic downturn of 2001-2002 appears to have constrained available resources, shrinking endowments and diminishing capacity for personal philanthropy.**

often linked to their interests in developing or exercising products. As another example, faculty at the California Institute of the Arts were involved for many years with Yamaha in the development of digital musical instruments, especially the piano, as a result of Yamaha's expectation that meeting the artistic needs of those musicians would result in a better commercial product. University of California at Los Angeles artist Bill Seaman's work on a "hybrid invention generator" is supported by a grant from the Intel Corporation. And as noted in Chapter 2, Ben Rubin at the Brooklyn Academy of Music benefited from Lucent Technologies and Rockefeller Foundation sponsorship for a collaborative project.

Within private philanthropy, data are most readily available for foundation activities. Grant support in 2000 was on the order of $3.7 billion for the arts, culture, media, and the humanities, a doubling of funding since 1996.[37] This growth was the result of a combination of factors, including a healthy economy, an increase in the number of new foundations, and significant increases in investments by major funders. Consequently, private foundations' share of all private giving to the arts increased from less than 30 percent to about 35 percent.[38] The economic downturn of 2001-2002, compounded by extraordinary demands for resources arising from the terrorist attacks of September 11, 2001, appears to have constrained available resources, shrinking endowments and programs among established foundations and diminishing capacity for personal philanthropy.[39]

As one might guess, the allocation of grant dollars is concentrated in particular areas of the United States. In a 1998 study conducted by the Foundation Center, five states—New York, California, Texas, Minnesota, and Michigan—and the District of Columbia received about 54 percent of art dollars and 52 percent of grants.[40] The same report showed a decrease in the share of grant dollars controlled by the top 50 recipients, which declined from 32.1 percent (or 7.9 percent of all arts grants) in 1992 to 28.5 percent (or 7.7 percent of all arts grants) in

---

[37]"The nation's nearly 56,600 grant-making foundations provided an estimated $3.69 billion for arts, culture, media, and the humanities in 2000, more than double the $1.83 billion estimated for 1996" (Renz, 2002, "The Foundation Center's 2002 Arts Funding Update," p. 1). Note that this $3.69 billion figure is compatible with the $10 billion figure cited earlier: The former represents grants awarded; the latter, total giving.

[38]Renz, 2002, "The Foundation Center's 2002 Arts Foundation Update," p. 1.

[39]For example, the David and Lucile Packard Foundation announced that it will no longer focus on arts and non-profit effectiveness, consolidating its activities in other areas after its endowment shrank from $13 billion in 1999 to $3.8 billion in mid-2002. See Jon Boudreau, "Packard Foundation Facing Cutbacks," September 19, 2002, available online at <http://www.bayarea.com/mld/mercurynews/4112039.htm>.

[40]Loren Renz and Steven Lawrence, 1998, *Arts Funding: An Update on Foundation Trends*, Third Ed., Foundation Center, New York, in cooperation with Grantmakers in the Arts, Seattle, p. 22.

1996.[41] The size of awards has also changed, with the number of smaller grants ($10,000 to $49,999) decreasing and mid-sized grants ($50,000 to $499,999) increasing, each by about 1 percent a year since 1996, with larger grants ($500,000+) remaining stable.[42] The median size of arts grants in 1999 was $25,000, the same as in 1992 and 1996 but slightly lower than the median amount for all foundation grants, which was $25,361.[43] Funding distribution by category in 2000 found the performing arts receiving 32.2 percent of arts-grants dollars, followed by museum activities (29.1 percent), media and communication (9.9 percent), and cross-disciplinary arts (8.8 percent).[44]

Foundations seem particularly interested in the role and expression of culture,[45] which seems less explicit in many of the government programs and has become more complex as a result of globalization and multiculturalism. These trends have required foundations to develop a broader understanding of the needs of arts groups and artists through more proactive consultations, seeking advice from the key figures from artistic domains, and prompting an increased reliance on research.[46] The resulting insights have led many foundations to experiment with new funding models, including venture capital; one-time endowments to start-up arts and cultural non-profits; and large, one-time grants. For example, the Ford Foundation's Education, Media, Arts and Culture program recently initiated the New Directions/New Donors for the Arts program, committing $42.5 million in challenge grants to 28 arts and cultural institutions and nearly quadrupling the foundation's annual arts appropriations.[47]

Foundation grants seem to be a significant source of support for generating, as well as providing access to, works of art, often emphasizing the value of art and cultural activities in building and strengthening communities. The AT&T Foundation, for example, emphasizes support for projects that "promote artistic expression or create net-

---

[41]"The disproportionate concentration of support among a relatively few recipients characterizes foundation funding in many fields. In 1996, the top 25 health recipients accounted for 29.4 percent of all health grants dollars, while the top 25 education recipients benefited from 22.3 percent of foundations' education dollars" (Renz and Lawrence, 1998, *Arts Funding*, p. 9).

[42]*A Snapshot: Foundation Grants to Arts and Culture*, 1999, Grantmakers in the Arts, Seattle, Wash.

[43]*A Snapshot: Foundation Grants to Arts and Culture*, 1999, Grantmakers in the Arts, Seattle, Wash., p. 4.

[44]Renz, 2002, "The Foundation Center's 2002 Arts Funding Update," p. 2.

[45]Arts and culture have been the fourth largest foundation funding priority since the mid-1980s; the first three are education, health, and human services. See Renz, 2002, "The Foundation Center's 2002 Arts Funding Update."

[46]Loren Renz and Steven Lawrence, 1999, *Arts Funding 2000*, Foundation Center, New York, in cooperation with Grantmakers in the Arts, Seattle.

[47]The program seeks to "link prosperity to creativity" by "offering opportunities for artists to work in new directions and play innovative roles in their communities." Grantees are encouraged to share best practices with each other and the arts community at large.

works that support artists and/or cultural organization" and engage local communities. A core program is AT&T: NEAT (New Experiments in Art & Technology), which provides support to regional science and children's museums to showcase work by artists who use science and/or technology as their creative medium.[48] The Rockefeller Foundation provides funding for a range of creative explorations in new media, including fellowships, conferences, publications, digital art exhibitions in museums and on the Web, and experimental laboratories that foster collaborations among artists, scientists, and technologists.[49] Since 1994, the Rockefeller Foundation has awarded over 100 grants in the area of new media.

Private philanthropy is more likely to help to sustain individual artists than is government support, which seems more oriented toward institutions. This is particularly true of smaller foundations. For example, the Media Arts program of the Jerome Foundation, which supports artists in New York City and Minnesota, focuses on providing grants to risky projects by individual emerging artists.[50] Many of the resources provided by private foundations for individual artists are channeled through non-profit intermediaries that are focused on a specific domain. Consider the Warhol Foundation as an example. It does not support individual artists directly, but rather gives to "cultural organizations that in turn, directly or indirectly, support artists and their work."[51] So, in 2002 the foundation provided a grant to Franklin Furnace Archive to support the production of live art over the Internet, indirectly funding the production of artwork by individual artists. Efforts of the Creative Capital Foundation are described in Box 8.1. In 1999, 3 percent (about $46 million) of larger arts grants from larger foundations supported professional development, which includes fellowships and residencies; internships; scholarships; and awards, prizes, and competitions. Some private support for individuals comes in the form of commissions, which have some similarities in practice to research grants because of their reflection of the character of the commissioning entity. Also, when museums do commission work, it appears to be driven by their interest in making new art forms accessible.[52]

> Emergent, small-scale, and experimental initiatives face a greater challenge in raising funds.

---

[48]See a description of the program at <http://www.att.com/foundation/programs/arts.html#tech>.

[49]A number of the individuals, organizations, projects, and conferences discussed in this report received funding from the Creativity and Culture Program of the Rockefeller Foundation. For further information, see <http://www.rockfound.org/display.asp?Context=3&SectionTypeID=16&Preview=0&ARCurrent=1>.

[50]See the Jerome Foundation's Web site at <http://www.jeromefdn.org/>.

[51]See the Warhol Foundation grant awards at <http://www.warholfoundation.org/FiscalYear2002F.htm>.

[52]For a discussion of commissioning practice, see Susan Morris, 2001, *Museums and New Media Art*, a research report commissioned by the Rockefeller Foundation (mimeo), October.

---

**BOX 8.1**

**Creative Capital Foundation**

An unusual approach to funding new arts projects is used by the Creative Capital Foundation.[1] Seen as an attempt to "fill the void left by the NEA [National Endowment for the Arts] when it stopped funding individual artists,"[2] Creative Capital was founded in 1999 as a non-profit organization in the state of New York. In its own words, Creative Capital seeks to "support artists creating original work who are pursuing innovative, experimental approaches to form and/or content in the visual, performing, and media arts."[3] Creative Capital also hopes to distinguish itself from traditional arts grant programs by providing marketing and other "non-artistic" aid (e.g., helping artists develop audiences for their work). Also, unlike most traditional funding sources, Creative Capital shares a portion of the proceeds generated by its artists' projects—if the projects indeed make money—to help replenish its funds and continue the cycle of support for future projects. One of the main areas where Creative Capital seeks to focus its attention and funds is emerging fields, which include, among other things, technology-based work. Indeed, early in 2000, Creative Capital was poised to deliver nearly $100,000 (or 16 percent of the foundation's total grants) to support 12 digital arts projects.[4] Other funding areas include new media, performing arts, and visual arts. By December 2001, Creative Capital had made grants to individual artists amounting to more than $1.5 million.[5] For the period from 1999 to 2002, more than 40 other foundations and individual donors provided almost $7 million for Creative Capital's programs and operations. Creative Capital's funders include, to name only a few, the Andy Warhol Foundation, Jerome Foundation, Rockefeller Foundation, Ford Foundation, Benton Foundation, and Home Box Office.

---

[1]See the Creative Capital Foundation's home page at <http://www.creative-capital.org/>.

[2]Shayna Samuels, 1999, "New Foundation Seeks Provocative Artists for Grants," *Dance Magazine* 73(8): 29.

[3]See <http://www.creative-capital.org/general/html/prospectus.html>.

[4]Matthew Mirapaul, 2000, "Digital Artists Draw Support from a New Foundation," *New York Times*, January 1, p. F14.

[5]Creative Capital, 2001, *Creative Capital: 2001 Year End Report*, Creative Capital, New York. Available online at <http://www.creative-capital.org/news&events/html/rubys_reports/2001YearEndReport.pdf>.

---

But even though it helps to sustain artists, private philanthropy tends to be associated more with grants to institutions than to individuals. Foundations and individuals often give to established institutions, which have name recognition; emergent, small-scale, and experimental initiatives face a greater challenge in raising funds. The largest shares of foundation arts grants go to program and capital support, which tend to be associated with institutions. The smallest share goes to research—a condition that must be improved to foster ITCP in the long run.

Among larger foundations, there appears to be some recognition of the importance of promoting creativity and expanding the uses of IT, but such foundations are just beginning to embrace IT as a theme.[53]

---

[53]See Computer Science and Telecommunications Board, National Research Council, 1998, *Advancing the Public Interest Through Knowledge and Distributed Intelligence (KDI)*, White Paper, available online at <http://www.cstb.org>.

They have been developing IT systems as part of their own infrastructure and as infrastructure for their grantees—but generally in programs other than those that support the arts, where infrastructure has not historically been a concern. This situation began to change in the late 1990s, when a number of the larger, most visible foundations began self-assessments and started new initiatives featuring some attention to IT. And, as the present project demonstrates, foundations have begun to contemplate the broad potential of ITCP.

For the most part, computer science research does not benefit from foundation support, unlike biomedical and some other forms of scientific research.[54] However, individuals and, in particular, corporate philanthropy support some academic IT research (in the form of grants for research, donation of computer hardware and software, or endowment of faculty positions). Corporate support, both in-kind and financial, is considered critical for growing and sustaining the relatively high cost of operations of university computer science departments. Corporate interactions are an important element of how those departments do business (e.g., personnel interactions, consultancies, job placement for students and graduates, and so on).[55] Hence computer scientists often emerge from their education with a relatively positive view of corporations as employers and sources of support, whereas artists will generally not have comparable experiences as an integral part of their education.

## Prizes

One notable source of funding for the arts is prizes (see Chapters 5 and 7).[56] Artists value prizes because they are expressly linked to

> Computer scientists often emerge from their education with a relatively positive view of corporations as employers and sources of support, whereas artists generally do not have comparable experiences as an integral part of their education.

---

[54]To put this observation into context, foundation grants for science and technology overall amounted to 3 percent of total dollars awarded in 2000, while grants for arts and culture amounted to 12 percent. See "Highlights of the Foundation Center's Foundation Giving Trends," 2002, Foundation Center, New York. Available online at <http://fdncenter.org/research/trends_analysis/>.

[55]Corporate conduct of research falls under the broader heading of commercial activity excluded from the scope of this discussion. Corporate IT research tends to be applied; regardless of label, it may be more development than research. See Computer Science and Telecommunications Board, National Research Council, 2000, Making IT Better.

[56]"Awards and prizes have proliferated throughout the arts, suggesting that numerous sponsors stand ready to fill the available ecological niches. The interests of these sponsors are evident enough. Associations of artists seek to advance professional standards and leverage their creative preferences against choices driven by the interests of humdrum participants and profit seekers. Independent philanthropists who sponsor and support prizes likely have the same objective. Prizes given by associations of critics serve to advertise and dignify the critics' regular services. Commercial sponsors seek goodwill for their products among the honorees and the public with more than casual interest in the relevant art world. . . ." See Richard E. Caves, 2000, Creative Industries: Contracts Between Art and Commerce, Harvard University Press, Cambridge, Mass., pp. 198-199.

critical appraisal and they provide visibility—a benefit to the entity awarding the prize as well as the recipient. As a Canadian arts council explains,

> Most organizations that provide assistance for the arts and culture have established or manage awards for artistic excellence accompanied by cash prizes. This practice has the advantage of combining the recognition of artists and their art with their contribution to cultural activity. Moreover, it significantly enhances the prestige of the role of funding agencies and publicly indicates that they significantly support artistic creation.[57]

Against this backdrop, prizes have begun to emerge for ITCP (see the section "Validation and Recognition Structures" in Chapter 7), building on the foundation laid by traditional supporters of the arts and economic development (e.g., various state and local prizes, or certain foreign ones, such as Europriz). Unlike grants, which may be multiyear or renewable, prizes are one-time infusions of revenue, and they are awarded ex post facto, which implies that the beneficiary needs some other source of support to do the work that competes for the prize.

## Federal Funding for Information Technology Research

It is well-known that the federal government has a long history of supporting IT research, which is largely computer science research (including computer engineering) plus some in sister fields such as electrical engineering. Indeed, it can be argued that federal support is at the very root of the information-technology world that Americans and others inhabit today.[58] Research funding supports the advancement of a field and associated knowledge, often because that field is linked to economic well-being and to the broad base of scientific and technical knowledge; it is also tied more specifically to meeting a government mission, which varies among agencies. Public access to information generated or published by an agency, like education, is also a concern to varying degrees among agencies.

Funding for IT research comes primarily from the Department of Defense, notably the Defense Advanced Research Projects Agency (DARPA), and from the National Science Foundation (NSF). A variety of other agencies also support computer science research, both funda-

---

[57]See Conseil des arts et des letters Québec, at <http://www.calq.gouv.qc.ca>.

[58]For more information on the history of federal support for computing research, as well as a history of the early Internet, see Computer Science and Telecommunications Board, National Research Council, 1999, *Funding a Revolution*, National Academy Press, Washington, D.C.

**BOX 8.2**

**Goals for the Networking and Information Technology Research and Development Program**

1. Ensure continued U.S. leadership in computing, information, and communications technologies to meet federal goals and to support U.S. 21st-century academic, industrial, and government interests.

2. Accelerate deployment of advanced and experimental information technologies to maintain world leadership in science, engineering, and mathematics; improve the quality of life; promote long-term economic growth; increase lifelong learning; protect the environment; harness information technology; and enhance national security.

3. Advance U.S. productivity and industrial competitiveness through long-term scientific and engineering research in computing, information, and communications technologies.

SOURCE: National Science and Technology Council, 2002, *Strengthening National, Homeland, and Economic Security: Networking and Information Technology Research and Development, FY 2003 Supplement to the President's Budget,* July.

mental and especially applied.[59] Computer science and related funding that is tracked in the aggregate as the Networking and Information Technology Research and Development (NITRD) program amounted to about $1.8 billion in 2002 (see Box 8.2 for NITRD goals).[60]

The NSF, which funds research across science and engineering disciplines, issues grants that tend to be smaller in size (tens to hundreds of thousands compared to hundreds of thousands into the millions) and duration (a few months or 1 to 2 years compared to 3 or more years) than those awarded by DARPA, which supports the development of large, complex, and often comparatively capital-intensive systems.[61] The NSF also supports research centers (for science

---

[59]The coordinated information technology research and development program suite tracked by the National Coordination Office is budgeted at between $1.5 billion and $2 billion. Other participating agencies besides NSF and various components of the Department of Defense include the Department of Energy (the Office of Science and the National Nuclear Security Administration), the National Aeronautics and Space Administration, the Department of Health and Human Services (the National Institutes of Health and the Agency for Health Care Research and Quality), the Department of Commerce (the National Institute of Standards and Technology and the National Oceanic and Atmospheric Administration), and the Environmental Protection Agency.

[60]Derived from National Science and Technology Council, 2002, *Strengthening National, Homeland, and Economic Security: Networking and Information Technology Research and Development, FY 2003 Supplement to the President's Budget,* July, the annual compendium of federally funded computer science research programs known informally as the "Blue Book."

[61]NSF has begun to experiment with support for larger computer science research projects. The early 2000s have witnessed intensive reexamination of program emphases at DARPA, with significant changes in its support for computer science research beginning to unfold but uncertain as of this writing.

and technology and for engineering research), which typically involve more than one discipline and sometimes multiple institutions. For example, the Integrated Media Systems Center at the University of Southern California[62] has undertaken research on new types of user interfaces with applications from everyday communication to interaction with art forms.

The terrorist attacks of September 11, 2001, have increased attention to security across the government (and industry), which is expected to influence trends in computer science research funding.[63] Projects and programs that can be linked to homeland security are expected to be favored, with uncertain implications for truly exploratory work for which no such linkage can be posited. Prior to September 11, the trend in the predecessor initiatives in what is now known as the NITRD program was a broadening in the scope of research, including more diverse, cross-disciplinary explorations that included the social sciences and humanities—an evolution toward greater attention to ITCP interests.

The evolution of federal funding programs for computer science has been associated with an erosion of flexibility. When computer science was a new field, in the late-middle 20th century, research successes were tied to a few talented individuals: program managers in government agencies, who selected researchers in universities and awarded money with considerable flexibility. That fabled pattern eroded considerably in the latter part of the century, as funding became tied more to specific government program objectives (which varied in their flexibility) and program management evolved to be more conservative overall, thanks to pressures on federal spending and increases in efforts to promote federal accountability. These circumstances militate against a program manager's experimentation with activities that can seem or be seen as new and different, such as ITCP—unless the linkage to other kinds of research and national benefits can be established and communicated effectively.

## FUNDING FOR INFRASTRUCTURE

Grants, regardless of source, fund both specific activities (e.g., research, development of new works of art) and the development of associated human resources, which can increase the likelihood of success and provide capacity for future work. Similar rationales drive support for relevant physical infrastructure, or tools, a critical aspect of work involving IT. Accordingly, ITCP work increases overall funding requirements on the arts side by introducing a need for hardware

> The evolution of federal funding programs for computer science has been associated with an erosion of flexibility.

---

[62]See <http://imsc.usc.edu/>.

[63]For example, the Blue Book sports the title *Strengthening National, Homeland, and Economic Security: Networking and Information Technology Research and Development, FY 2003 Supplement to the President's Budget* for the edition published in mid-2002.

and software infrastructure already provided for computer science.[64] This can add significant costs to budgets.[65] However, dramatic improvements in IT price/performance ratios—making this infrastructure less expensive than it was just a few years ago—have enabled smaller groups or individuals to do things that were prohibitively expensive for them in the past. Grant support may be particularly important in building the ITCP infrastructure inasmuch as IT firms are less likely to provide hardware and software for activities seen as part of the (non-profit) arts world, tending instead to donate such components for more conventional R&D.[66] There are exceptions where marketing publicity may be gained; some artists have served as beta testers for IT products and have allowed the use of their names in marketing in return. As noted above, there is already a history of foundation support for capital projects, the category under which IT infrastructure is likely to fall.

One productive approach to infrastructure building, illustrated by the federal Digital Libraries Initiative (DLI) (see Box 8.3)—is the application testbed concept, which is familiar to experimental computer scientists. Another approach, illustrated by some DLI projects and a wide range of other computer science and humanities research projects, is the generation of databases and other information resources that are made available to the public and could be exploited for ITCP. There is already evidence of artists drawing on government repositories of imagery, for example.

Whereas a testbed is associated with developing something, ITCP also presents needs for infrastructure associated with archiving and preservation, without which too much of it may prove ephemeral. It is far too often the case that ITCP-based works disappear with the close of the specific, original project. Even ITCP work of distinction, which has been recognized and awarded prizes, often meets that fate when

---

ITCP

presents needs

for infrastructure

associated with

archiving and

preservation,

without which too

much of it may

prove ephemeral.

---

[64]Computer scientists, in turn, tend to complain that research funding budgets, by aggregating support for research per se with support for infrastructure, which often benefits other kinds of scientists, overstates what is available to them.

[65]State and regional economic development that features local investment in networking infrastructure can help educational and cultural institutions. It can also benefit individuals through initiatives that promote deployment of residential broadband, for example (see Computer Science and Telecommunications Board, National Research Council, 2002, *Broadband: Bringing Home the Bits,* National Academy Press, Washington, D.C.). On a national scale, the Internet2 project seeks to connect a growing number of colleges and universities to comparatively high-bandwidth/capacity networking to support research and education and long-distance collaboration. The cyber infrastructure exploration at NSF, drawing on multiple initiatives, also links dispersed researchers with a range of networked resources. Depending on where practitioners are located, ITCP can benefit from—and provide added motivation for—such extended infrastructure investments.

[66]IT firms have long recognized that giving their products to educators and researchers helps to develop their customer base. The arts communities seem to lack that kind of appeal for firms, though there are exceptions (e.g., Apple Computer has seeded art, music, and architecture departments to be identified with creative users).

---

**BOX 8.3**

**Awards Related to ITCP Under the Digital Libraries Initiative**

*A Live Performance Simulation System: Virtual Vaudeville ($900,000 over 3 years)*

The Virtual Vaudeville project brings together a diverse array of scholars, including computer scientists, three-dimensional modelers and animators, theater practitioners, and theater and music historians. The objective is to use digital technology to address a problem fundamental to performance scholarship and pedagogy: How to represent and communicate the phenomenon of live performance using media. This problem becomes especially pressing when the objective is to represent a performance tradition from the past. Neither a written description nor a filmed re-creation can convey the experience of attending a live performance, an experience that encompasses not only the way the performance on stage looks and sounds from different parts of the theatre, but also spectators' perceptions of, and interactions with, one another. The proposed solution to this problem is to re-create historical performances in a virtual reality environment. The central objective is to simulate a feeling of "liveness" in this environment: the sensation of being surrounded by human activity onstage, in the audience, and backstage, and the ability to choose where to look at any given time (onstage or off) and to move within the environment.

*Capturing, Coordinating, and Remembering Human Experience ($200,000 over 4 years)*

This work will develop algorithms and systems enabling people to query and communicate with a synthesized record of human experiences derived from individual perspectives captured during selected personal and group activities. For this research, an experience is defined through what you see, what you hear, and where you are, and associated sensor data and electronic communications. The research will transform this record into a meaningful, accessible information resource, available contemporaneously and retrospectively. This vision will be validated through two socially relevant applications: (1) providing memory aids as a personal prosthetic or behavioral monitor for the elderly, and (2) coordinating emergency response activity in disaster scenarios.

---

SOURCE: Adapted from <http://www.dli2.nsf.gov/itsprojects.html>.

---

the financial and/or institutional support dries up. The creators of such products are sometimes desperate to secure a further application for their work but have no means to carry this idea through to fruition (e.g., a multimedia kiosk presentation accompanying an exhibition could well be used for education purposes after the closure of the exhibition for which the presentation was created). Thus, works that are not seen as having commercial value will only be made available in digital form by the non-profit and government sectors that allocate resources to do so. Large-scale digital conversion is likely to incur significant financial costs, which most non-profits are not in a position to absorb.[67]

---

[67]See commissioned papers of the Art, Technology, and Intellectual Property project of the American Assembly, available online at <http://www.americanassembly.org/ac/atip_p_cp.htm>.

## RISK PREFERENCES AND THE CHALLENGE OF SUPPORTING EMERGING AREAS

Grants are usually awarded in the context of formal programs (government and private), which define the parameters of activities suitable for funding. Depending on the source, there will be more or less discretion to support activity that does not fit easily within the stated parameters; sometimes there is an explicit intent to fund opportunistic new ideas, whereas other times there is strict adherence to guidelines, with no flexibility. Program emphases change at different rates and in different directions. In computer science, for example, funding for supercomputing, graphics, artificial intelligence, and other subdisciplines has waxed and waned over time with perceptions of need and opportunity.

The risk preferences of grant makers affect what gets funded; they also contribute to a greater flow of resources to institutions than to independent workers. The early DARPA, mentioned above, is an oft-cited example of grant makers willing to fund risky endeavors; the Information Sciences Program of a mid-20th-century sister agency, the Office of Naval Research, just happened to be led by an individual with strong arts interests, which had some effect on funding directions. A tendency toward risk aversion among grant makers can militate against movement into new areas or new types of work processes. Risk aversion can arise for many reasons, chief among them the fear of failure and the need to carry out fiduciary responsibilities, because any award has an opportunity cost—money allocated to one recipient is not available to others. On the government side, risk aversion may arise in response to growing pressures for accountability in spending. For example, the Government Performance and Results Act (Public Law 103-62), although tricky to apply to research, requires agencies to link projects they support to the results achieved. This is easier to do for applied research than for more exploratory, fundamental research, where some degree of failure is to be expected, as is unanticipated and possibly delayed success. In addition to precluding some adventurous research, risk aversion among grant makers can lead to constraints on grantees, such as requirements for more frequent (and therefore disruptive) reporting or demonstration of work in progress.[68]

Of greatest concern for ITCP as an emerging arena for research is risk aversion related to the selection of topics. Some examples are obvious: In an environment with a new, major focus on homeland security and counterterrorism, projects that seem to involve interfering with or destroying a system might not be viewed favorably. Young researchers in computer science and engineering, who either served on or briefed the committee, reported a need to package ITCP work in more conventional terms to increase the likelihood of funding. Al-

> Of greatest concern for ITCP as an emerging arena for research is risk aversion related to the selection of topics.

---

[68]This has been a growing complaint among computer science researchers.

though it has always been true that grant seekers have to use care in presenting their ideas, this phenomenon can limit the exploration of a new arena such as ITCP by discouraging applications for funding or eliminating ideas that do not lend themselves to repackaging. Absent grant program definitions that specifically embrace ITCP, progress in ITCP will depend on grant seekers' ingenuity in influencing program definitions and relating their ideas to an existing definition.[69]

The "culture wars" of the 1990s underscore that reactions to artistic endeavors can have political overtones and consequences. These external concerns can influence which arts projects are funded, whereas influences on funding in computer science research tend to depend on such criteria as relevance to the operational mission of an agency (NSF is an exception) or the objectives of a given program. Program definitions and scopes may be more or less inviting of the risk involved in new approaches, and so may the processes of evaluating competing proposals (although the nominal criteria, such as NSF's "intellectual merit" and "broader impacts," may not make the potential for variation obvious). On the arts side, there is anecdotal evidence that focus and selection processes can discourage innovation, especially at the point when new art forms emerge. At present, these tensions shape the ITCP arena of digital arts or new-media work, which may be treated as a separate category or recognized as a part of other fields, depending on the funding source. Funders are trying to understand what is emerging and the nature of its merits as a contender for their resources. The IT elements have engendered mixed reactions among traditional arts funders, just as arts elements have perplexed traditional computer science funders.

Risk aversion also arises among foundations. Whereas government agencies have some degree of public oversight and accountability, foundations are intrinsically idiosyncratic, shaped by the preferences of their founders and boards of directors. Those preferences may constrain the program staff's solicitations of and reactions to proposals.

Recently, federal programs have sought to promote cross-disciplinary work related to IT—albeit in areas other than ITCP. This situation has put agency leadership, with grants as the proverbial motivational carrots, somewhat at odds with the research community: Grant seekers have not, on the whole, gravitated easily to cross-disci-

---

[69]One speaker at a committee meeting suggested that there might be limited funding available for a proposal on "computer music" research, but much more funding available if the proposal were relabeled as "digital signal processing" or "digital audio signal" research. A Washington, D.C., briefing to the Computer Science and Telecommunications Board in September 2002 by Chuck Thorpe of Carnegie Mellon University presented some pragmatic contrasts: Research involving robotic soccer-playing dogs could be seen as illustrating dogs playing or as multiagent collaboration in a hostile situation. In other instances, principal investigators have skirted innovation by simply renaming old proposals to obtain funds.

plinary collaboration. Whether impelled by government programs or arising through a bottom-up process from the grant-seeking community, the relative paucity of cross-disciplinary work to date means that the researchers serving on the panels that review proposals may lack experience in this type of work and/or an appreciation for its merits and potential. Anecdotal reports from committee members and grant seekers and program managers who briefed the committee suggest that these panels are most comfortable with projects that come closest to core disciplinary work, or are unable to select for quality among competing proposals with differing approaches to cross-disciplinarity, especially where new or unfamiliar methodology may be involved. This situation is not unique to ITCP; in any context, selection panels may have difficulty addressing the reinterpretation of fields or emergence of new fields, inasmuch as it involves work and people diverging from panel members' own experience. The problem is well-known among federal program managers.

The NSF (with some collaboration with other government organizations) has engaged in three notable experiments in supporting cross-disciplinary work with an IT component, some of which relates to ITCP. (See Box 8.4.) The first two, Knowledge and Distributed Intelligence (KDI) and the Digital Libraries Initiative (mentioned above), featured strong internal champions at NSF, who worked with the relevant research communities; the third, Information Technology Research (ITR), drew from a broad, cross-agency reconceptualization of federal programs for IT R&D, also with input from the research community. KDI's difficulties reflected its attempt to achieve the broad involvement of multiple disciplines; the other two may have benefited in terms of program viability from their greater leaning toward computer science (i.e., a single discipline). The integration of social sciences with computer science is an element of all three, but DLI also moved to involve humanists, and some ITR components are defined in ways that are open to the involvement of humanists and artists. These and other cross-cutting research initiatives, through both substantive and procedural emphases, aim to advance the science and practice of collaboration as a means to achieving creativity in any context, another factor that makes them relevant to planning for the support of ITCP.

The committee emphasizes that investigating new areas and exerting leadership are risky, and therefore grant makers need to develop a tolerance for some failures as the inevitable price of exploration. Understandably, foundations prefer that all projects succeed, because they feel a responsibility to maintain trust both as a fiduciary body and as a legacy of a family, company, or community. There is a tendency to play it safe. A way to institutionalize risk into the grant-making process is to allocate a fixed amount or percentage of grants for risk-taking projects—initiatives with potentially large payoffs but also with a significant possibility of failure. This is grant making as portfolio investing: combining safe, slow-growth projects with some riskier choices, and expecting some failures as normal. For this

*Investigating new areas and exerting leadership are risky, and grant makers need to develop a tolerance for some failures as the inevitable price of exploration.*

## BOX 8.4
### Cross-disciplinary Computer Science Initiatives Pointing Toward ITCP

*Knowledge and Distributed Intelligence*

Knowledge and Distributed Intelligence (KDI) was an experimental initiative carried out between 1997 and 1999 that cut across the National Science Foundation (NSF) directorates (defined by their emphases on research in the physical, natural, and social sciences, and education) and confederated multiple programs (such as the Digital Libraries Initiative discussed below, Learning and Intelligent Systems, Knowledge Networking, Universal Access, and Integrated Spatial Information Systems). The initiative revolved around the relationship between research on information technologies and efforts to meet societal needs, broadly defined.

Because of the initiative's sweeping vision, program management featured not only coordination across units of NSF but also outreach to other agencies (those involved in specific component programs) and private foundations. Both program managers and the research community grappled with objectives and options through formative workshops and other interactions, but a lack of consensus about the initiative and skepticism among groups of researchers (generally defined along disciplinary boundaries) about each other's roles and their integration weakened the experiment. KDI did not extend beyond its initial 3-year phase, although component programs have continued.

*Digital Libraries Initiative*

The Digital Libraries Initiative (DLI) has grown to embrace a widening set of federal agencies, research disciplines, and application contexts. Its first phase (launched in 1994) involved six universities funded by the NSF that developed testbeds and engaged a variety of partners in diverse applications domains. Its second phase, DLI2 (beginning in 1998), which has added relevance to ITCP, expanded to include as partners the National Endowment for the Humanities, Library of Congress, Defense Advanced Research Projects Agency, National Library of Medicine, and National Aeronautics and Space Administration. The DLI generates infrastructure in applications domains that provide a context for developing new computer science—hence the basing of the NSF participation in the Computer and Information Science and Engineering (CISE) directorate. Collaboration between computer scientists and very different communities is intrinsic to DLI. Also see Box 8.3 and <http://www.dli2.nsf.gov>.

*Information Technology Research*

Information Technology Research (ITR) is an umbrella initiative for a wide range of computer science research projects, including core disciplinary investigations and research that engages other disciplines to varying degrees. Based at NSF, it reflects that organization's approach to a larger reconceptualization of support for IT research across several agencies. Like that larger effort, ITR has been widening its scope from fundamental computer science to computational science and to interactions with multiple disciplines—the fiscal year 2002 funding was supposed to emphasize "emerging opportunities at the interfaces between information technology and other disciplines" that can "elucidate, expand and exploit IT."[1] ITR addresses social, economic, and workforce implications, involving social scientists to address how to design computer-based systems that work better for people, as well as the societal impacts of systems as currently designed (see, for example, NSF's Digital Society and Technology program).[2] The engagement of social scientists involves both the CISE and the Social, Behavioral, and Economic Sciences directorates.

---

[1]See <http://www.itr.nsf.gov/aboutitr.html>.

[2]See <http://www.interact.nsf.gov/cise/descriptions.nsf/30ff6e7ea7d05a0d85256659004c0237/976ffb01fa3aea5e852565d90056864d?OpenDocument>.

approach to work, program managers must not be subject to sanctions for the inevitable failures. In general, tolerance for occasional failure may be higher in the private sector than in the public sector; hence, private foundations may have to take the leadership with respect to high-risk initiatives.

Federal funding agencies have special issues to consider. To begin with, the agency structures and approaches typical of the IT arena—science and engineering research—differ significantly from those used in the arts and humanities. The level and allocation of funding, nature of funding agencies, interactions among agencies, and so on, are generally different, although there are some commonalities. Experience suggests that the success of any new major programs within the U.S. government will be enhanced by the appointment of a champion. The structure of the federal government—in both the executive and congressional branches, the latter strongly influenced by the appropriations process—can often militate against cross-disciplinary programs cutting across jurisdictional and disciplinary lines. Without a champion, no one understands the potential benefits or who is accountable and whose budget is at stake.

When funding a major new program in ITCP, organizations may increase their chances for success by following general principles for fruitful research and "good work." Funding should be made available to cross-disciplinary teams—perhaps a dozen teams or so are needed to attain critical mass, in the committee's judgment—and the budget should be adequate to support two or more principal investigators. The program should be "owned" by several units within a foundation or government agency (or multiple foundations or agencies, or possibly even a combination thereof) to obtain real, committed participation by all. Commitments should be made for a sufficient period to enable programs to become established—on the order of 5 years—and allow for no (or very limited) provision for renewals, to avert intentional or accidental empire-building.[70]

## REEXAMINING FUNDING POLICIES AND PRACTICES

Reliance on grant support (so-called "soft" money) shapes the modus operandi of academic researchers and artists, chiefly by building the search for grants into their activities. A characteristic of grant support is finite duration: Even longer-term grants build in the anticipation of an end to the support. Also, with the possible exception of support from a research center, individual grants tend to provide only partial support for a computer science researcher or an artist. As a result, professionals who depend on grant support work to develop

---

[70]Based on Norman Metzger and Richard Zare, 1999, "Interdisciplinary Research, "From Belief to Reality," *Science* 283: 642-643.

and renew grants, a process with uncertain results that typically involves attempts to seek more grants than one is likely to receive. The awarding of longer-term grants allows the grant makers to get to know the grant seekers better. Inasmuch as grant seeking is a competitive process, participants may benefit from having a track record. But officials at federal funding agencies also understand that young researchers may need to be evaluated somewhat differently than are their senior counterparts.[71] It is not clear whether foundations are comparably sensitive to that problem. Committee members seemed to regard foundations as conservative and therefore most likely to fund those with a track record. Some specific disciplines, such as architecture, are structured with the expectation that solo creativity can flower comparatively late in a career, and job opportunities are structured accordingly—but architecture operates within a market context.

The issue of whether to fund "research" or "content" also arises; these concepts, which never could be separated with complete clarity, are becoming more intertwined.[72] Grants that foster ITCP are particularly likely to incorporate both content and IT in substantive ways. Ad hoc practices, such as pairing content producers with researchers to qualify them for research funding, can be awkward and inefficient—working through the challenges of collaborations (see Chapter 2), but only for economic reasons. Instead, granting agencies should review their policies for supporting content development, or consider cooperative projects among consortia of granting agencies in areas of special interest, to meet challenges that cannot otherwise be addressed. On the arts side, private philanthropic collaborations have already begun to nurture ITCP. For example, the Rockefeller Foundation has joined forces with the Ford Foundation and the Pew Charitable Trusts in exploring relevant issues.

One interesting experiment is a grant awarded to Leonardo/ISAST to study the feasibility of a hybrid art center and research lab structured to be financially sustainable. Called Arts Lab, the project is an attempt to build a bridge between the creative community exploring new technologies and the marketplace. Arts Lab is structured as a not-for-profit corporation that is intended to be managed with the discipline of a commercial enterprise. The goal is to be sustainable with little compromise of artistic or research values. This type of project can help in exploring the boundary between commercial and non-commercial spheres: Arts grants, in particular, have been assumed to support non-commercial activity, but given the nature of ITCP and the need to promote viability, this boundary bears reexamination.

Agencies and foundations should structure proposal-review processes that encourage the development of new practices as well as the evolution of existing ones. As has happened previously in various scientific and engineering disciplines, digital practices are proliferat-

---

[71]NSF, for example, offers career awards to young investigators.
[72]This is one reason for the prominence of design in ITCP.

ing within all existing categories of the arts, from filmmaking and theater to the fine arts, as well as in new categories. Should the digital arts be placed in their own category or recognized as a part of everything else? Both strategies should be pursued to expand existing definitions as well as recognize new and emerging forms. Accordingly, a wide range of representation is necessary on panels, and panelists will need to be familiar with multiple areas—a long-standing issue concerning appraisal panels and the peer-review process. As part of this effort to nurture innovation, panels should consider making, in lieu of only large grants, some modest grants to a larger number of groups or individuals. To avoid forcing individuals or small organizations to spend disproportionate amounts of time on grant-related paperwork, care may be needed in program design, or implementation strategies could be pursued that leverage intermediaries that can absorb some of the administrative burdens.[73]

Multiple approaches are important, because ITCP is evolving at a rate that seems to be outrunning the traditional feedback loop of criticism, and it is evolving in multiple directions—there is no consensus on how to appraise different ideas. How funders react will shape whether and how schools of thought or practice develop. To fund a diversity of projects, program managers will need more time and leeway to learn about, evaluate, and reach out to accommodate increased numbers of grantees. They need competence to reinterpret existing fields to accommodate the evolution associated with ITCP and to place priority on truly new forms of creative practice that exploit the power of digital information and networks. They may also need to convene different groups of evaluators who are peers to different kinds of grant seekers and also have an appreciation for how fields and creative practices are changing.

Funders are receiving feedback from their decision-making panels that indicates it is difficult to review the full spectrum of practice now offered within certain granting categories. For instance, a media panel might have to judge a documentary film alongside an interactive installation or Web site. Panelists often feel unable to judge both effectively. How can agencies structure a panel process that will encourage the development of new practices as well as the evolution of existing ones? In other words, how can agencies support development without defining it before it has emerged? One problem can be that people with experience and credentials in a field may feel threatened or discomfited by the onslaught of new media forms. Selection processes need to cover the breadth of evolving practices without balkanizing panels and without slanting choices to the particular expertise represented. It will be important to find ways to function as support for the incubators rather than as a prescriptive reinforcement of existing and familiar forms of work.

> Program managers will need more time and leeway to learn about, evaluate, and reach out to accommodate increased numbers of grantees.

---

[73]The National Alliance of Media Arts and Culture (<http://www.namac.org>) might be an appropriate model for a national umbrella organization to distribute funding to local community media arts centers.

Support for development of new practices also entails recasting what is regarded as research, to enable expanding activities in organizations (notably government technical research funders) with comparatively little history in funding artistic practices and activities in organizations (notably arts branches of foundations) with comparatively little history in funding IT research. Québec-based funding agencies, for example, have identified an entirely new realm of supportable activity defined by the term "research-creation," defining a practice that straddles the artistic and the technical and does not carry some of the constraints of technical research. This approach helps to reduce the chances of projects falling through the cracks in a world where, for example, projects are not funded by the NEA because they are perceived as technical research, but also are not funded by the NSF because they are seen as cultural projects. As discussed in Chapter 3, there would be value in establishing a special granting category for tool building in both software and hardware. Additional support would be needed to make these tools available (i.e., distribution, documentation, and support) to both artists and industry and generally push forward creative production.

As noted above, archiving and preservation concerns also have to be addressed by funders. The instigators and funding bodies of ITCP work should feel responsible for the long-term preservation and use of this work whenever possible (see discussion of archiving in Chapter 7). The need for long-term archiving and/or preservation can be addressed both through infrastructure support and through augmentation of the creativity-oriented grant itself.

Getting at the IT-arts intersection involves attention to both the scope of work and the mechanisms of combining different disciplines, whether through the efforts of an individual or cross-disciplinary teams. By drawing on multiple sources of funds—within a complex organization like the NSF or across organizations—a program might reinforce the message that a cross-disciplinary effort is sought, although that raises its own practical problems at the program management end.[74]

Experimentation with different approaches to grant making would correspond to the novelty and dynamism of ITCP. One model of note is the Pew Charitable Trusts Venture Fund. Created during the mid-1990s to enable the Trusts to explore opportunities that fall outside its traditional program areas, the Venture Fund has no restrictions on the subject matter it can support. Key funding decisions include such questions as, Are we being innovative? Are we maximizing our return on investment? Answers to these questions are evaluated based on social returns, not financial ones. Other questions include the follow-

> By drawing on multiple sources of funds, a program might reinforce the message that a cross-disciplinary effort is sought.

---

[74]The challenges of combining funding among private foundations and between NSF and private foundations are outlined in the CSTB white paper *Advancing the Public Interest Through Knowledge and Distributed Intelligence (KDI)*, 1998.

ing: Does the proposed project have a particular urgency? Is it addressing an important need? Can it produce concrete results in a reasonable time frame? Will it enable us to explore new partnerships with other foundations or with public or private funders? If it carries a significant risk of failure, is there also the possibility of unusually high social return if it succeeds? During 2001 the fund issued more than $34 million to cover 20 grants. Since its inception, the fund has provided several grants to organizations associated with information technology and creativity. For example, it joined the Rockefeller Foundation in providing support to the American Assembly's Art, Technology and Intellectual Property (ATIP) project addressing the impacts, challenges, and opportunities resulting from technological advances that are confronting the arts.[75]

Informed analysis of the related issues will require increased evidence gathering and analytical resources for the arts and humanities.[76] Relatively few economists, social scientists, or policy analysts engage in policy research related to the arts and humanities (as compared to other domains), in part because of ambivalence in the United States about the public role in this space.[77] The empirical base for supporting policy analysis is weak, and the academic and practitioner wings of the community are largely strangers. Interaction among policy makers, practitioners, and policy scholars is much richer in other fields of public policy.[78] In addition to building a knowledge base in this area, public policy analysts could build relationships and understanding with key intermediaries, such as journalists, commentators, think tank researchers, and political party representatives. Policy attention to these issues is consistent with the long-run shift of the economy toward services, some of which focus on cultural products and content generally, and enduring concern about the quality of life and creativity in general. Finally, and complementary to developing better data, a digital art and culture history project would provide valuable context

> Relatively few economists, social scientists, or policy analysts engage in policy research related to the arts and humanities. The empirical base for supporting policy analysis is weak.

[75]See <http://www.pewtrusts.com> and <http://www.americanassembly.org/ac>.

[76]See Ruth Ann Stewart and Catherine C. Galley, 2002, "The Research and Information Infrastructure for Cultural Policy: A Consideration of Models for the United States," appendix in J. Mark Schuster, *Informing Cultural Policy: The Research and Information Infrastructure*, Rutgers Center for Urban Policy Research, Center for Urban Policy Research Press, New Brunswick, New Jersey.

[77]A consortium of foundations moved to address this practical problem by underwriting the Center for Art and Culture (see <http://www.culturalpolicy.org/issuepages/infotemplate.cfm?page=History>). The Pew Charitable Trusts made news by voicing concerns about "cultural policy," which is more controversial in the United States than in many other countries. Its initiative "Optimizing America's Cultural Policies" announced in 1999 was renamed "Optimizing America's Cultural Resources" in response to the ensuing controversy over the appropriateness of this kind of policy. See Schuster, 2002, *Informing Cultural Policy*.

[78]See Margaret Jane Wyszomirski, 1995, "Policy Communities and Policy Influence: Securing a Government Role in Cultural Policy for the Twenty-First Century," *Journal of Arts Management, Law, and Society* 25(3): 192-205.

and serve as an educational tool for policy makers, educational institutions, and people interested in engaging in ITCP.[79]

• • • • • • • • • • • • • • • • • • • • • • • • • • • • • • • • • • • • • •

# FUNDING IN THE INTERNATIONAL CONTEXT

Internationally, a variety of ITCP funding models have evolved. Examples are presented here both to illustrate the diversity of approaches and to draw contrasts between the United States and other countries. As noted earlier, a nation's cultural policies influence funding for art and design activities. A case could be made that a stronger central intervention (e.g., the creation of a ministry of culture, as in France) can have positive effects, either to get new ITCP initiatives going, or at least jump started, and/or to help establish an infrastructure. (Funding sources may be reluctant to pay for an infrastructure until there is a demonstrated need for it; but the need may not materialize until the infrastructure is in place.) However, a case could equally be made that decentralized funding fosters initiatives that rise from the bottom up and thus are more likely to lead to the development of ideas and projects that reflect leading-edge work in ITCP.

## Public Support for the Arts

Worldwide, funding for the arts has a long history[80] that has been far from uniform across time or space. Indeed, looking primarily at European countries, one is likely to form the opinion that direct government support for the arts is a long-standing and widespread practice. Looking at other countries such as the United States or Japan, however, one might just as easily conclude that national governments leave support for the arts primarily to private (commercial or nonprofit) or local organizations. Kevin Mulcahy[81] provides a succinct

---

[79]For further discussion, see Kevin F. McCarthy and Elizabeth Heneghan Ondaatje, 2002, *From Celluloid to Cyberspace: The Media Arts and the Changing Arts World*, RAND, Santa Monica, Calif.

[80]Public support for the arts among the nations of the world is almost as old as civilization itself. Indeed, examples of such support in even the distant past are fairly easy to find; one need only consult an art history or world history text to find numerous instances. Pisistratus (605?–527 B.C.), for instance, although known as the "tyrant of Athens," is also remembered as a patron of the arts. He arranged city-state support for a range of artistic activities, including building projects, poetry, sculpture, dance, and music. He is also credited with starting public arts festivals that were open to all citizens in an effort to enhance the cultural prestige of Athens.

[81]See Kevin V. Mulcahy, 1998, "Cultural Patronage in Comparative Perspective: Public Support for the Arts in France, Germany, Norway, and Canada," *Journal of Arts Management, Law and Society* 27(4): 247-264.

description of the perceived differences in how Europe and the United States, in particular, have dealt with public arts support historically:

> The conventional wisdom of much discourse about public support of the arts is that European national governments are long-time, generous, and uncritical benefactors of culture, whereas the U.S. government, by invidious comparison, has been a reluctant supporter, of decreasing generosity and with increasingly dispiriting criticism. But broad generalizations about comparative public policies often disguise substantial exceptions.[82]

The developing nations present another, much more resource-constrained picture, but as with economically stronger nations they, too, support—or receive support from such non-governmental organizations as foundations for—the arts in the context of cultural heritage and competitive advantage. In developing nations, the traditions of art and aesthetics have developed high degrees of sophistication whose potential for interaction with IT has barely been explored. Information technology, when combined with the arts and crafts in these countries, can accelerate economic and human development. This can take simple forms, such as the use of IT to create markets for the creative output of the populations in these nations. It can also take the form of inspiration for new designs rooted in the aesthetics and traditions of the local culture.[83] Both could leverage financial support, to the extent it is available. See Box 8.5.

France has a long and rich tradition of public support for arts and culture, dating back at least to the Capet monarchy (c. 987). The French government is among the world's largest funders of art and culture, administered primarily through the Ministry of Culture and Communication.[84] Created in 1959, the ministry seeks to make art and culture available to as much of the French public as possible, with an annual budget of nearly 1 percent of the entire national budget.[85] A specific division of the ministry of culture and communications in France finances research and development. A major item included in this

---

[82]The substantial exceptions derive primarily from the indirect public support provided by donors motivated or rewarded (at least in part) by U.S. income tax incentives, as described in the previous section. However, though constrained by available data sources, a comparison of direct government arts funding suggests a huge variation per capita, ranging from $6 in the United States to $46 in Canada and $85 in Germany. Data derived from National Endowment for the Arts, 2000, *International Data on Government Spending on the Arts*, Research Division Note #74, available online at <http://arts.endow.gov/pub/Notes/74.pdf>.

[83]Ranjit Makkuni's work is an example; see <http://www.hinduonnet.com/thehindu/2001/04/23/stories/13230074.htm>.

[84]See the ministry's home page at <http://www.culture.fr>.

[85]One percent of the U.S. federal executive branch budget for FY 2002 is approximately $13 billion.

---

**BOX 8.5**
**Four Models for Public Arts Funding**

• The *facilitator state* supports the arts through foregone taxes, which is to say that donors' contributions are made tax deductible. The objective of this model is to promote diversity of activity in the non-profit amateur and fine arts, although "no specific standards of art are supported" by the state. Rather, the focus is on "the preferences and tastes of the corporate, foundation, and individual donors." The United States is a good example of a facilitator state.

• The *patron state* provides support for the arts through arm's-length arts councils. In this model, the state decides on an overall level of support, leaving the actual decisions regarding which projects to support to the arts councils. The arts councils, in turn, rely on the advice of professional artists working through a system of peer evaluation. The objective of this model is as much to promote "standards of professional artistic excellence" as it is to support "the process of creativity." The United Kingdom is an example of a patron state.

• The *architect state* provides funds for the arts through government institutions created solely for that purpose (e.g., ministries or departments of culture or the arts). In this model, the state tends to support the arts as part of its national objectives, which can have distinctively different emphases; for example, historically, France has leaned toward professional standards of artistic excellence, whereas the emphasis in the Netherlands is much less elitist.

• The *engineer state* "owns all the means of artistic production." In this model, the state supports only art that meets certain "political standards," and decisions about funding and support are left up to "political commissars." One example of an engineer state was the Soviet Union.

---

SOURCE: Adapted from Harry Hillman-Chartrand and Claire McCaughey, 1989, "The Arm's Length Principle and the Arts: An International Perspective—Past, Present and Future," *Who's to Pay for the Arts? The International Search for Models of Support*, M.C. Cummings, Jr. and J. Mark Davidson Schuster, eds., American Council for the Arts, New York. Available online at <http://www.culturaleconomics.atfreeweb.com/arm's.htm>.

---

budget is the Institut de Recherche et Coordination en Acoustique et Musique (IRCAM), a permanent state-funded institute for computer music research. This office also offers "State Commissions" to French artists, and on occasion these commissions have been opportunities for international co-production (e.g., the virtual Tunnel under the Atlantic, "constructed" between Paris and Montréal in September 1995). Major institutes such as IRCAM and Germany's Zentrum für Kunst und Medien (ZKM) described below (and in Chapter 5) derived the basis for political support of major public expenditures in part to respond to U.S. high-technology leadership.[86]

The German method for directing public funds to the arts is much more decentralized. Indeed, the German federal government has been described as "effectively barred" from cultural activities, a response to earlier times when government-supported cultural activities were used

---

[86]IRCAM's efforts to combine high modernist musical experimentalism with technology transfer to the French IT sector were fraught with contradictions and confusion, according to one ethnographic study. See Georgina Born, 1995, *Rationalizing Culture: IRCAM, Boulez, and the Institutionalization of the Musical Avant-garde*, University of California Press, Berkeley.

"for purposes of national glorification," or even "abused . . . for propaganda purposes."[87] Accordingly, Germany has no central agency to oversee cultural funding; rather, individual länder (or states) and local authorities bear this responsibility. For example, although many of ZKM's constituencies are outside Germany, much of its funding is derived from regional public funds and small-scale industrial sponsorship. Such a unique environment comes with a high price tag, and to contribute to the institution's discretionary funds, the research institutes also secure contracts for ITCP research within European Union (EU) research projects, which has caused tension between the institutes' mission to develop artwork and the need to generate funding through research projects that generally do not directly fund art as such. Local governments account for 47 to 58 percent of public spending on the arts, the states for 40 percent, and the federal government for only 2 to 13 percent; despite tax incentives, private support for the arts accounts for "no more than 1 percent" of the total revenue for German art institutions.[88] Various networks of research institutes (e.g., the Fraunhofer institutes) may have entities whose work relates to ITCP. The Fraunhofer, for example, even has an outpost in Rhode Island (near Brown University and the Rhode Island School of Design) to support its interests in computer graphics.

Government programs established to support the arts have histories measured in decades rather than centuries in many countries. For example, the Canadian government's main means of supporting the work of individual artists and arts organizations is the Canada Council for the Arts, which was created by an act of Parliament in 1957 and was funded initially by a $50 million endowment to "ensure the Council's complete independence of the government."[89] Currently, however, in addition to support from various endowments, donations, and bequests, the council receives the majority of its funding from Parliament in the form of an annual appropriation. In 2000-2001, the Canada Council made awards and grants amounting to $117 million;[90] in April 2002, the council and the National Research Council (NRC) of Canada launched a collaborative program of research grants to bring leading artists into Canadian NRC laboratories across the country as researchers.[91] This agreement will also create a forum to facilitate the development of partnerships among other arts, science, and technology organizations in Canada. In addition, the Canada Council will soon complete its first round of awards to joint proposals

[87]Annette Zimmer and Stefan Toepler, 1996, "Cultural Policies and the Welfare State: The Cases of Sweden, Germany, and the United States," *Journal of Arts Management, Law, and Society* 26(3): 167-195.

[88]See Zimmer and Toepler, 1996, "Cultural Policies and the Welfare State."

[89]See John Meisel and Jean Van Loon, 1987, "Cultivating the Bushgarden: Cultural Policy in Canada," in Milton C. Cummings, Jr., and Richard S. Katz, eds., *The Patron State: Government and the Arts in Europe, North America, and Japan*, Oxford University Press, New York, p. 289.

[90]From <http://www.canadacouncil.ca/council/about-e.asp>.

[91]See <http://www.canadacouncil.ca/news/pressreleases/co0215-e.asp>.

by groups of artists, engineers, or scientists, which will be reviewed for the first time by a mixed panel cooperatively managed by the arts council and its two sister science councils (the Natural Science and Engineering Council and the Social Science and Humanities Council). The government of Canada also underwrites the National Film Board, an innovative force in computer graphics (see Box 6.3 in Chapter 6), and the Canadian Broadcasting Corporation and offers subsidies, tax incentives, and marketing support to creative industries (publishing, music, museums, multimedia, computer games).

The Japanese experience with public arts support bears only a vague resemblance to that of Europe, Canada, or the United States. Currently, most art genres in Japan—including traditional Japanese arts, modern arts derived from Europe, and popular arts—are "thoroughly commercial," surviving through a mixture of income from, among other sources, ticket sales, advertising, and the sale of related products or services.[92] The flurry of Japanese corporate sponsorship for art and technology that began in the late 1980s was typically justified as a sophisticated kind of symbiotic corporate philanthropy (which has suffered in the difficult economic climate in Japan; see Chapter 5). However, the Japanese government has built museums and libraries and preserved important cultural monuments ever since the Meiji state (a stronger, more centralized government) was established in 1868.[93] Indeed, the Tokyo Academy of Music was founded in 1879, and the Academy of Art was formed in 1887; these two organizations merged following World War II to form the Tokyo University of Fine Arts, an institution considered to be the center of art and music research and education in Japan.

In 1968 the Agency for Cultural Affairs (ACA), devoted to public support for the arts, was formed and began with an initial budget of around $14 million, but over the next 10 years the ACA budget swelled, as authorities discovered that the promotion of culture was in the public's interest at home "as well as the national interest abroad."[94] In 1972, the Japan Foundation was created by a legislative act to be an autonomous non-profit public corporation whose primary concern is cultural relations abroad. It promotes a variety of cultural exchange programs each year, with its overall focus being on personal exchange. In 2002, the Association for Corporate Support of the Arts (Mecenat) broadened its definition of an arts event—in the context of authorizing tax-deductible contributions—to include media arts.[95]

---

[92]Meisel and Van Loon, 1987, "Cultivating the Bushgarden," p. 333.

[93]Thomas R.H. Havens, 1987, "Government and the Arts in Contemporary Japan," in Cummings and Katz, eds., *The Patron State: Government and the Arts in Europe, North America, and Japan*, p. 334.

[94]Havens, 1987, "Government and the Arts in Comtemporary Japan," p. 333. During the period from 1968 to 1978, the ACA budget grew some 574 percent, while the overall national budget increased by 489 percent.

[95]Dramatic Online, "Broadening the Definition of Arts Events in Japan," available online at <http://www.dramaticonline.com/ifacca/web/news/detail.asp?Id=24073&from=Arts_Council_News>.

## Public Support for Information Technology Research

Information technology research is supported by a number of national governments around the world. As in the case of the arts, the policies and practices of only a few countries are reviewed in this report, to provide a broader perspective on, and contrast to, policies and practices in the United States.

Canada uses a model that combines tax and funding incentives while also supporting collaboration and information sharing among geographically dispersed federal research labs, private R&D facilities, and research universities. A large part of this model is made possible as a result of the Canadian Network for the Advancement of Research, Industry and Education (CANARIE),[96] which is Canada's advanced Internet development organization. This non-profit organization, which receives its core funding from the Canadian government, has established the world's largest and fastest national R&D network. A collaboration involving more than 120 universities and industry partners, CANARIE has helped to fund more than $600 million in research projects related to the Internet, including those associated with content distribution.[97]

In addition to the tax incentives open to all forms of industrial R&D (in contrast to the United States[98]), the Canadian government funds a series of research organizations whose focus is promoting innovation in IT. The Canadian NRC operates the Institute for Information Technology (IIT),[99] which, through cost-sharing collaborative projects, assists other organizations with the development of market-driven technologies. These collaborations can be one-on-one with a single company or multiparty, with several participating organizations combining resources to share costs and risks. A similar program, the Networks of Centres of Excellence (NCE),[100] promotes partner-

---

[96]See <http://www.canarie.ca>. Also mentioned in Chapter 5.

[97]As of this writing the CANARIE project, *e-content*, is soliciting requests for projects that emphasize "cultural research, development and applications in areas such as architecture and design, film and video, 3D graphics, net and web art, digital music, digital photography, game design, graphic design, human/computer interface, and copyright/rights management tools; and feasibility studies, including consumer research/testing of broadband cultural products and the monetization of content." See <http://www.canarie.ca>.

[98]See *Canada's Leadership in Information and Communications Technologies*, available online at <http://strategis.ic.gc.ca/SSG/it04270e.html>.

[99]See <http://www.iit.nrc.ca/>.

[100]Three Canadian federal granting agencies, the Canadian Institutes of Health Research, the Natural Sciences and Engineering Research Council of Canada, and the Social Sciences and Humanities Research Council of Canada, along with Industry Canada, have combined their efforts to support and oversee the NCE. See <http://www.nce.gc.ca/>.

ships among industry, universities, and government. The program, which is intellectually diverse and geographically dispersed, consists of 22 centers, 8 of which conduct research in information and communication technology (usually, networks are funded for 7 years and can be renewed).[101] Since 1997, several proposals were made to the NCE program for hybrid research networks (grouping university-based researchers in IT, the arts and humanities, and cultural organizations (like the Banff Centre and the National Film Board)). Although none of these proposals passed the stringent NCE competition, they formed the basis for an alternative program with a regional basis (i.e., supporting geographic clusters) and support for new-media industrial activity. Canadian Heritage, the branch of the Canadian government responsible for cultural affairs, launched a new grant program in the fall of 2002 as part of its Canadian Culture Online program. Called New Media Research Networks, it was planned to support multiyear collaborative networks on a pilot basis through March 2004, with funding for up to five networks for 3 years for about one million Canadian dollars per year. Its goals include "development of an environment that is conducive to Canada becoming a world leader in digital context creation and production" through "research at the intersection of technologies and culture."[102]

Following a model similar to the NCE program is Precarn Incorporated,[103] a national consortium of corporations, research institutes, and government partners that supports innovation in intelligent systems. Precarn helps to promote collaboration among Canadian companies, universities, and government researchers by providing funding on a case-by-case basis for projects that include the participation of at least two companies and one university.

Japan is known for a strong national, coordinated policy with respect to technology, and the situation with IT is no different. The policy strategy for IT that the Prime Minister, Council on Science and Technology Policy (CSTP), and the ministries currently support is the Basic Law on the Formation of an Advanced Information and Telecommunications Network Society, or e-Japan for short. The primary goal of e-Japan is to "establish an environment where the private sector, based on market forces, can exert its full potential and make Japan the world's most advanced IT nation within five years."[104] E-Japan is an integrated strategy that emphasizes four points:

---

[101]For example, the Centre of Information Technology and Complex Systems involves more than 50 university researchers supported by 379 graduate students, and carries out research in seven Canadian provinces. The network supporting this center includes 28 universities, 62 industrial partners, and 27 government departments and agencies. See *Canada's Leadership in Information and Communications Technologies*, available online at <http://strategis.ic.gc.ca/SSG/it04270e.html>.

[102]See <http://www.canadianheritage.gc.ca/progs/pcce-ccop/progs/mednet_e.cfm> and <http://www.pch.cg.ca/progs/pcce-ccop/pubs/mednetguide_e.cfm>.

[103]See <http://www.precarn.ca/>.

[104]See *e-Japan Strategy* at <http://www.kantei.go.jp/foreign/it/network/0122full_e.html>.

- Enable every citizen to enjoy the benefits of IT.
- Reform the economic structure and strengthen industrial competitiveness.
- Realize affluent national life and creative community with vitality.
- Contribute to the formation of an advanced information and telecommunication networked society on a global scale.

Under the CSTP, a Cabinet-level office sets science and technology (S&T) policy direction and funding levels. The CSTP uses a comprehensive overview strategy in its decision making that has resulted in a significant increase in the involvement of the humanities and social science in the discussion. Increasingly, this strategy is forcing policies into a direction that emphasizes the relationship between society and human beings.[105] The CSTP immediately identified the support of IT as one of four strategic fields that the government must emphasize when formulating S&T policy and funding R&D. However, while the CSTP decides on policy direction and funding levels, it does not fund individual projects directly; this is left to the ministries. Government funding for IT R&D comes primarily from three ministries: the Ministry of Economy, Trade, and Industry (METI);[106] the Ministry of Education, Culture, Sports, Science and Technology (MEXT);[107] and the Ministry of Public Management, Home Affairs, Posts, and Telecommunications.[108] Specifically for IT R&D, the government has developed an approach whereby it will promote market competition, facilitate commercialization, and promote international cooperation and collaboration among industry, academia, and government agencies.[109]

A strong, coordinated position has also been taken in Europe. Most recently, as part of its Fifth Framework Programme for Research, Technological Development and Demonstration Activities (FP5) (1999–2002), the European Commission identified seven thematic research programs intended to promote industrial competitiveness and quality of life in Europe. One of these, the Information Society Technologies (IST) Programme, is designed to support R&D in information and communication technologies. The IST Programme had a working budget of 3.6 billion euros (1998–2002) to achieve its goal of supporting IT research "within a single and integrated programme that reflects the convergence of information processing, communication, and media

---

[105]See *A New System for Promoting Science and Technology in Japan*, available online at <http://www.nsftokyo.org/rm01-15.html>.

[106]Formerly known as the Ministry of International Trade and Industry. METI oversees about 15 percent of the R&D budget.

[107]The former Ministry of Education, Science, Sports, and Culture and the Science and Technology Agency have been combined to form MEXT, which oversees about 63 percent of the S&T budget.

[108]Formerly two separate ministries.

[109]See "Overview of Action Plan" at <http://www.kantei.go.jp/foreign/it/990519overview.html>.

technologies."[110] The European Commission manages the program with the assistance of the IST Committee, consisting of representatives of each EU member and associated states. A key feature of the program is its emphasis on supporting cross-program (CP) themes. The objective of CP projects is to allow grant seekers the flexibility to address cross-disciplinary research and ensure that topics associated with more than one area are addressed. For example, one CP project, Technology Platforms for Cultural & Arts Creative Expressions, is concerned with "developing future generic platforms and tools for improving creative expression and facilitating access to inspirational material for artistic and cultural content creation."[111] To achieve this goal, the IST seeks to support medium- and long-term exploratory work with an emphasis on digital expression by providing funding to collaborative projects.

As the European Commission continues to develop its next framework (6), the IST Programme remains one of the main themes and a significant part of the anticipated S&T needs. Specifically, within the Sixth Framework, information and communication technologies are being looked at to "stimulate the development in Europe of technologies and applications at the heart of the creation of the Information Society in order to increase the competitiveness of European industry and allow European citizens in all EU regions the possibility of benefiting fully from the development of the knowledge-based economy."[112] To achieve this goal, the research supported by the IST Programme will focus on "the future generations of technologies in which computers and networks will be integrated into the everyday environment . . . that places the user, the individual, at the center of the future developments for an inclusive knowledge-based society for all."[113] Some concerns have arisen, however, about the adequacy of support to be provided for ITCP research and development. The RADICAL consortium—three European media arts and cultural organizations engaged in a 2-year project—issued a manifesto in July 2002 urging non-market support for transdisciplinary ITCP activity. Its recommendation that "specific support mechanisms be implemented to promote cross-disciplinary research platforms which explicitly include media arts and cultural organisations and creative practitioner-researchers," in the context of the rest of the document, raises questions about the depth and durability of official support for ITCP in Europe.[114]

---

[110]See "Background" at <http:europa.eu.int/information_society/programmes/research/index_en.htm>.

[111]See "Objectives" at <http://www.cordis.lu/ist/cpt/2002cpa15.htm>.

[112]See "IST in FP6, Priority" at <http://www.cordis.lu/ist/fp6/fp6.htm>.

[113]See "IST in FP6, Priority" at <http://www.cordis.lu/ist/fp6/fp6.htm>.

[114]See "The RADICAL Manifesto" at <http://www.e-c-b.net/ecb/internal/1027149839>. The RADICAL consortium includes the SMARTlab at Central Saint Martins College of Art and Design in the United Kingdom, the Ecole supérieure de l'image in Angoulême-Poitiers in France, and the Society for Old and New Media in the Netherlands, together with a network of artists, creative professionals, and small and large businesses participating in the RADICAL program of events and symposia.

## PRIVATE PHILANTHROPY

Outside the United States, private philanthropy seems to play an important, although relatively less prominent, role as a source of funding for the arts and IT. Although the committee did not attempt a comprehensive survey of relevant programs, one initiative did come to its attention that deserves mention. The Daniel Langlois Foundation has established the Program for Organizations from Emerging Regions,[115] through which it financially supports projects that allow artists or scholars who are not European or North American to immerse themselves in technological contexts that are non-existent or difficult to access in their own country. The aim of the program is to promote the integration of knowledge and practices specific to different cultures, and grants may also support research projects that combine traditional artistic practices with advanced technology or that explore methods and processes based on the unique aesthetic principles of certain cultures. Each year the foundation selects two different priority regions. Examples of projects funded through the program include the following:

• *Lima, Peru: Alta Technologia Andina (ATA), Media Laboratory for Education, Research and Creation in Video and New Media.* Funding was used to help integrate electronic elements into local artistic practices and to set up a shared space for cross-disciplinary education and the creation and distribution of experimental artwork.

• *Delhi, India: Center for the Study of Developing Societies: Interface Zone.* Funding was used to create a physical meeting place in Delhi to act as a dynamic node for fostering the exhibition, online dissemination, and pedagogy of new-media culture.

• *Sofia, Bulgaria: InterSpace Media Arts Center.* Funding was used to support new-media projects and to showcase projects generated by the media lab in an effort to foster sustainable growth in the independent artistic scene in Bulgaria and to encourage artistic experimentation with new media.

• *Riga, Latvia: The Center for New Media Culture: Acoustic Space Research Lab and Program.* Funds were used to establish two media labs and to develop the Acoustic Space Research Program to investigate the field of streaming media; to coordinate projects in sound art, audio, radio, and streaming media; and to organize international events on sound and acoustics.

---

[115]The primary source of information for this overview was the Daniel Langlois Foundation Web site at <http://www.fondation-langlois.org>. For 2002, the foundation gave priority to West Africa and South America.

# Appendixes

# A | Biographies of Committee Members and Staff

**WILLIAM J. MITCHELL,** *Chair,* is professor of architecture and media arts and sciences and dean of the School of Architecture and Planning at the Massachusetts Institute of Technology (MIT). He has demonstrated an unusual interest in how technology and society interact, team-teaching with software innovator Mitch Kapor and spearheading the development of a new program within the School of Architecture/Department of Urban Studies and Planning in that area. He has developed software and engaged in distance learning, and he oversees the MIT Media Laboratory, an alternative (to the mainstream) concentration of information technology expertise at MIT. Mitchell teaches courses and conducts research in design theory, computer applications in architecture and urban design, and imaging and image synthesis. He consults extensively in the field of computer-aided design and was the co-founder of a California software company. He has served recently on the Council for the Arts and Technology at Central State University (Ohio). Mitchell came to MIT in 1992 from the Graduate School of Design at Harvard University. From 1970 to 1986, he was on the faculty of the School of Architecture and Planning at the University of California at Los Angeles. He has also taught at Yale, Carnegie Mellon, and Cambridge Universities. Mitchell holds a B.A. from Melbourne University, a master of environmental design from Yale University, and an M.A. from Cambridge University. Among his many writings is *City of Bits.*

**STEVEN ABRAMS** manages the Business and Application Modeling group in the Software Technology Department at IBM Research. With that team, he researches and develops tools that help people describe, architect, visualize, validate, and develop enterprise applications more easily and naturally than traditional tools. He originally joined IBM in 1992 to develop two-dimensional and three-dimensional geometric processing algorithms and system architecture for a rapid prototyping system in the Manufacturing Research group, while pursuing his Ph.D. He then joined Stratasys Inc., where he helped to commercialize the rapid prototyping technology developed at IBM. After finishing his Ph.D., he rejoined IBM Research, working in the Computer Music Center. As manager of that department, he led a series of projects that revolve around music, art, and creativity. One goal of the work of the Computer Music Center was to develop a better understanding of how people work on creative tasks in general, and how technology can better support people in these tasks. In his new role, Abrams is taking lessons learned from the music domain and applying them to the creative tasks of designing, architecting, and developing software systems. Abrams studied at Columbia University, where he earned B.S. and M.S. degrees in computer science from the School of Engineering and Applied Sciences in 1990 and 1991, respectively, and the M. Phil. and Ph.D. degrees from the Graduate School of Arts and Sciences in 1993 and 1997, respectively. His Ph.D. thesis was on sensor planning for robots in an active environment, focusing on multidimensional modeling and manipulation of computer vision constraints and the computation of three-dimensional swept volumes.

**MICHAEL CENTURY** is chair of the Arts Department at Rensselaer Polytechnic Institute, which he joined in August 2002. Long associated with the Banff Centre for the Arts, Century founded the Centre's Media Arts Division in 1988 and instigated the Art and Virtual Environments project (1991-1994). From 1993 to 1996, Century was a program manager at the Canadian Centre for Information Technology Innovation (CITI), a federal research laboratory located in Montréal, with responsibility for new-media arts funding. From 1996 to 1998, he served as policy advisor to the federal department of Canadian Heritage. Since September 1997, he has been the principal of Next Century Consultants, focusing on new media and cultural policy for various public and university sector clients. From 1997 to 2001 he was a research fellow and adjunct professor at the Graduate Program in Communications, McGill University, Montréal. For the Rockefeller Foundation, he researched and wrote a report in 1999 entitled *Pathways to Innovation in Digital Culture*. He was educated in humanities, piano performance, and musicology at the University of Toronto (B.A.), the University of California at Berkeley (M.A.), and the University of Iowa (M.A). He has recently completed a historical study of the transition from analog to digital techniques in animation as a doctoral dissertation in science and technology policy studies at the University of Sussex.

**JAMES P. CRUTCHFIELD** received his B.A. in physics and mathematics in 1979 and his Ph.D. in physics in 1983 from the University of California at Santa Cruz. He is currently a research professor at the Santa Fe Institute (SFI) after 14 years in the Department of Physics at the University of California at Berkeley (UCB). He maintains a research group at SFI that includes postdoctoral researchers and Ph.D. students. Currently he is an adjunct professor of physics in the Physics Department at the University of New Mexico, Albuquerque. He was a visiting research professor at the Sloan Center for Theoretical Neurobiology at the University of California, San Francisco, a postdoctoral fellow of the Miller Institute for Basic Research in Science at UCB, a UCB Physics Department IBM postdoctoral fellow in condensed matter physics, a distinguished visiting research professor of the Beckman Institute at the University of Illinois, Urbana-Champaign, and a Bernard Osher Fellow at the San Francisco Exploratorium. At the exploratorium, he helped design and mount the "Turbulent Landscapes" exhibit series on chaos, complexity, and pattern formation. This exhibit series was funded by the National Science Foundation and Department of Energy and was on display from June 1996 through January 1997. It is currently touring the nation's science museums. Crutchfield has several publications, his most recent being *Quantum Automata and Quantum Grammars, Theoretical Computer Science* (2000).

**CHRISTOPHER CSIKSZENTMIHALYI** is a member of the faculty at the Massachusetts Institute of Technology Media Laboratory. Previously, he was an assistant professor of electronic art at Rensselaer Polytechnic Institute. He has worked in the intersection of new technologies, media, and the arts for 8 years, lecturing, showing new-media work, and presenting installations in both Europe and North America. His most recent piece, Natural Language Processor, was commissioned by the KIASMA Museum in Helsinki, Finland. He has an M.F.A. from the University of California at San Diego (1998) and a B.F.A. from the School of the Art Institute of Chicago.

**ROGER DANNENBERG** is a senior research computer scientist on the faculty of Carnegie Mellon University's School of Computer Science. Dannenberg's current work includes research on music understanding, the automated accompaniment of live musicians, and the design and implementation of high-level languages and systems for real-time control and signal processing. His current artistic direction is toward real-time integrated computer music and computer graphics performance systems, for which he has developed tools for rapid software prototyping. Dannenberg has a broad background in electrical engineering and computer science. His publications include work on computer music, human-computer interaction, program verification, programming language design, computer architecture, and computer networks. Dannenberg is also a musician and composer. He performs frequently on trumpet in classical, jazz, and contemporary ensembles and writes works for electronic and conventional media. Dannenberg earned his Ph.D. (1982) and an M.S. in computer science from Carnegie Mellon University, as well as an M.S. in computer engineering from Case Western Reserve University. He also has a B.S. degree in electrical engineering from Rice University.

**TONI DOVE** is an artist who works primarily with electronic media, including virtual reality and interactive video laser disk installations that engage viewers in responsive and immersive narrative environments. Her work has been presented in the United States, Europe, and Canada, as well as in print and on radio and television. Her most recently completed project, Artificial Changelings (1993-1998), is an interactive narrative installation that uses video motion sensing to track the location and movements of a viewer standing in front of a large screen and translates them into changes in image and sound. A sci-fi romance about shopping, this interactive movie follows the life of Arathusa, a kleptomaniac in 19th-century Paris during the rise of the department store, who is dreaming about Zilith, an encryption hacker in the future with a mission. It debuted at the Rotterdam Film Festival in 1998 and was part of the exhibition "Body Mécanique" at the Wexner Center for the Arts and at the Computing Commons Gallery at Arizona State University during the Performance Studies International Conference in March 2000. A new piece currently under development, Spectropia, will be a feature-length interactive movie. It will be performed by two players on multiple screens for an audience in a theatrical setting or experienced by two individuals interacting at the same time as a single-screen serial installation in three parts over a more extended time period. The second phase of the proposal is to port the project to a component-based DVD system for a shared interactive narrative experience in the living room. A research fellowship from the Institute for Studies in the Arts at Arizona State University will provide the programming and engineering resources to develop the technology prototype.

**N. KATHERINE HAYLES**, professor of English and media arts at the University of California at Los Angeles, teaches and writes on the relationships of science, technology, and literature in the 20th century. Her most recent book, *How We Became Posthuman: Virtual Bodies in Cybernetics, Literature and Informatics* (1999), won the Rene Wellek Prize for the best book in literary theory. Her current projects include two new books, *Coding the Signifier: Rethinking Semiosis from the Telegraph to the Computer* and *Linking Bodies: Hypertext Fiction in Print and New Media*. Her work has been recognized by a Guggenheim Fellowship, two fellowships from the National Endowment for the Humanities, a Rockefeller Residential Fellowship in Bellagio, and numerous prizes and awards, including the Distinguished Scholar Medal from the University of Rochester and the Medal of Honor from Helsinki University.

**J.C. HERZ** (jc@joysticknation.com) is the principal of Joystick Nation Inc., a research and design practice that applies the principles of complex systems to the design of products, services, and brands. Drawing from an understanding of ecology, online social dynamics, computer games, and information theory, Joystick Nation's focus is multiplayer interaction design and systems that leverage the intrinsic characteristics of networked communication. Clients include multinational corporations, high-tech start-ups, and military research organizations. She is the author of two books, *Surfing on the Internet* (Little Brown, 1994), an ethnography of cyberspace before the Web, and *Joystick Nation: How Videogames Ate Our Quarters, Won Our Hearts, and Rewired Our Minds* (Little, Brown, 1997), a history of video games that traces the cultural and technological evolution of the first medium that was born digital, and explores how it shaped the minds of a generation weaned on Atari. Herz published 100 essays on the grammar and syntax of game design in the *New York Times* between 1998 and 2000.[1] She has conducted workshops on game design and learning and has spoken at technology and design conferences, including Technology Entertainment Design (TED) in Monterey, SIGGRAPH, E3, Game Developers' Conference, and the Forum on the Future of Higher Education at the Aspen Institute. Herz sits on the Defense Advanced Research Projects Agency's study group on patterns of emergent behavior in massively multiplayer persistent worlds.

**NATALIE JEREMIJENKO** is a design engineer and techno-artist. Recently, she was named one of the top 100 young innovators by the MIT *Technology Review*, her work was featured in the Tate Gallery Cream 2, and a large project was commissioned for the opening of the museum MASSMoCA (<www.massmoca.org>). Her work includes digital, electromechanical, and interactive systems in addition to biotechnological work and has appeared in the Rotterdam Film Festival (2000), the Guggenheim Museum, New York (1999), the Museum Moderne Kunst, Frankfurt, the LUX Gallery, London (1999), the Whitney Biennial '97, Documenta '97, and Prix Ars Electronica '96, presented at the Museum of Modern Art in New York. She was a 1999 Rockefeller fellow. She did graduate studies at Stanford University in mechanical engineering and at the University of Melbourne in the History and Philosophy of Science Department, and her Ph.D. is in the Department of Information Technology and Electrical Engineering, University of Queensland. As the director of the Engineering Design Studio at Yale University she is developing and implementing new courses in technological innovation. She is also affiliated with the Media Research Lab/Center for Advanced Technology in the Computer Science Department at New York University, where she did postdoctoral studies. Other research positions include several years at Xerox PARC in the computer science lab and a position at the Advanced Computer Graphics Lab, RMIT University. Jeremijenko has also been on the faculty in digital media and computer art at the School of Visual Art, New York, and the San Francisco Art Institute. She is known to work for the Bureau of Inverse Technology.

---

[1] Available online at <www.nytimes.com/library/tech/reference/indexgame theory.html>.

**JOHN MAEDA**, Sony Career Development Professor of Media Arts and Sciences, is an associate professor of design and computation at the Massachusetts Institute of Technology (MIT) Media Laboratory. Maeda attended MIT, where he was awarded bachelor's and master's degrees in computer science in 1989. He completed his doctoral studies in graphic design at the Tsukuba University Institute of Art and Design in Tsukuba, Japan. There, he began to experiment with ideas on ways to bond the simplicity of good graphic design together with the complex nature of the computer. Those experiments grew into a series of five books, called *Reactive Books,* that are a worldwide-recognized standard for high-quality digital media design. His commercial work for Shiseido Cosmetics, Sony, and Morisawa was honored in 1996 in the one-man exhibition "John Maeda: Paper and Computer" at the Ginza Graphic Gallery in Tokyo, Japan, and at the Dai Nippon Duo Dojima Gallery in Osaka, Japan. In 1999, he produced *Design by Numbers* (MIT Press), which outlines the theoretical underpinnings of his work as a combination of graphical examples and codes. His latest book, *MAEDA@MEDIA* (Thames & Hudson/Rizzoli, 2000), outlines his design and technology philosophy. John Maeda's awards include the 1994 Japan Multimedia Grand Prix for "The Reactive Square," the 1996 Tokyo Type Director's Club Gold Prize for his series of 10 posters for Morisawa, the 1997 Tokyo Type Director's Club Interactive Prize for "12 o'clocks," the 1999 Japan Ministry of Culture Interactive Prize for "one-line.com," the 1999 ID Magazine Gold Prize, the 1999 Milia d'Or nomination, and the 1999 New York Art Director's Club New Media Gold Award for "Tap, Type, Write." He is also a 1999 recipient of the Daimler-Chrysler Design Award. He is an honorary member of the Tokyo Type Director's Club and is on the national board of directors of the American Institute of Graphic Arts.

**DAVID SALESIN** is a professor in the Department of Computer Science and Engineering at the University of Washington, where he has been on the faculty since 1992, and a senior researcher at Microsoft Research, where he has also worked since 1999. He received his Sc.B. from Brown University in 1983 and his Ph.D. from Stanford University in 1991. From 1983 to 1987, he worked at Lucasfilm and Pixar, where he contributed computer animation for the Academy Award-winning short film *Tin Toy* and the feature-length film *Young Sherlock Holmes.* During his years at Stanford, he also worked as an intern at the DEC Systems Research Center and Paris Research Lab. In 1991-1992, he spent a year on leave as a visiting assistant professor in the Program of Computer Graphics at Cornell University. He has consulted at Sogitec Audiovisual, Aldus (now part of Adobe), Xerox PARC, Broderbund, and Microsoft Research. In 1996, he co-founded two start-up companies, where he served as chief scientist: Inklination and Numinous Technologies (acquired by Microsoft in 1999). Salesin received an NSF Young Investigator Award in 1993; an ONR Young Investigator Award, an Alfred P. Sloan Research Fellowship, and an

NSF Presidential Faculty Fellow Award in 1995; the University of Washington Award for Outstanding Faculty Achievement in the College of Engineering in 1996; the University of Washington Distinguished Teaching Award in 1997; the Carnegie Foundation for the Advancement of Teaching and the Council for the Advancement and Support of Education 1998-1999 Washington Professor of the Year Award in 1998; and the ACM SIGGRAPH Computer Graphics Achievement Award in 2000. Salesin's research interests are in computer graphics and include non-photo realistic rendering, image-based rendering, color reproduction, digital typography, and compositing.

**LILLIAN F. SCHWARTZ** is best known for her pioneering work in the use of computers for what has since become known as computer-generated art and computer-aided art analysis, including graphics, film, video, animation, special effects, virtual reality, and multimedia. Her work was recognized for its aesthetic success and was the first in this medium to be acquired by the Museum of Modern Art. Her contributions in starting a new field of endeavor in the arts, art analysis, and the field of virtual reality have recently earned her Computerworld Smithsonian Awards. Schwartz began her computer art career as an offshoot of her merger of art and technology, which culminated in the selection of her kinetic sculpture, Proxima Centauri, by the Museum of Modern Art for its epoch-making "1968 Machine Exhibition." She then expanded her work into the computer area, becoming a consultant at the AT&T Bell Laboratories, IBM's Thomas J. Watson Research Laboratory, and Lucent Technologies, Bell Labs Innovations. On her own, and with leading scientists, engineers, physicists, and psychologists, she developed effective techniques for the use of the computer in film and animation. Besides establishing computer art as a viable field of endeavor, Schwartz additionally contributed to scientific research areas such as visual and color perception and sound. Her own personal efforts have led to the use of the computer in the philosophy of art, whereby databases containing information as to palettes and structures of paintings, sculptures, and graphics by artists such as Picasso and Matisse are used by Schwartz to analyze the choices of those artists and to investigate the creative process itself. Her contributions to electronic art analysis and restoration, specifically in Italian Renaissance painting and fresco, have been recognized. Schwartz's work has been much in demand internationally both by museums and festivals. Schwartz has always had close ties to the academic community, having been a visiting member in the computer science departments and psychology departments of several universities and colleges. She has also been awarded numerous fellowships and honors. There are several books that include her work, the most recent being *The Web* (2000), by Bridget Mintz Testa. She has also written, with Laurens Schwartz, *The Computer Artist's Handbook* (1992). For other publications, awards, lectures, collections, exhibitions, films, and videos, see <http://www.lillian.com>.

**PHOEBE SENGERS** is an assistant professor in the Computing and Information Science Department and in Science and Technology Studies at Cornell University. Her work synthesizes cultural studies and computer science, by building new technology based on a cultural critique of existing research practices and assumptions. Results of this work are evident in three areas: technical advances in artificial intelligence, human-computer interaction, and media research; cultural analyses of technical practices; and explorations of strategies for cross-disciplinary synthesis between the two cultures of the humanities/arts and the sciences/technology. Sengers graduated in 1998 from Carnegie Mellon University with a self-defined cross-disciplinary Ph.D. in artificial intelligence and cultural theory. In 1998-1999, she was a Fulbright guest researcher at the Center for Art and Media Technology (ZKM) in Karlsruhe, Germany. From 1999 to 2001, she was a research scientist in Media Arts Research Studies at the German National Research Center for Information Technology (GMD). In 1999, *Lingua Franca* named her one of the top 20 researchers most likely to change the way we think about technology.

**BARBARA STAFFORD** is the William B. Ogden Distinguished Service Professor at the University of Chicago in the Department of Art History. Her special interests include the relationships of art, science, and medicine in the early-modern period, and the history of perception, visualization, and the intellectual and cultural development of body imagery. She also works on contemporary media and visualization technologies. Her focus is on the intersections between the arts and sciences in the early-modern and modern periods. She writes contemporary art criticism and serves as a visiting critic. Professor Stafford is continuing her work on analogy and neurobiology and presented an exhibition on visionary technologies at the Getty Museum, Los Angeles, from November 2000 through February 2001. Stafford has written several books, including *Visual Analogy: Consciousness as the Art of Connecting* (MIT Press, 1999), *Good Looking* (1996), and *Essays on the Virtue of Images* (1996). She has also written several articles, including "A Range of Critical Perspectives: Digital Imagery and the Practices of Art History," *Arts Education Policy Review* (July-August 1998), and "To Collage or E-Collage?" *Harvard Design Magazine* (Fall 1998). Stafford has also delivered several lectures, her most recent being "Old Mind/New Mind: The Role of Images in the Consciousness Debates" (Wonder Conference, Santa Barbara, March 2000). In May 1998 she spoke on computers and writing at the University of Florida, Gainesville, and in April 1998 she was a speaker in the Jurassic Technology series at the School of the Art Institute of Chicago. Stafford received her Ph.D. in art history from the University of Chicago, as well as an M.A. in art history and a B.A. in philosophy and comparative literature from Northwestern University.

*Staff*

**ALAN S. INOUYE** is the study director for the Committee on Information Technology and Creativity, which created *Beyond Productivity: Information Technology, Innovation, and Creativity*, and is a senior program officer at the Computer Science and Telecommunications Board (CSTB). Inouye has a wide range of interests at the intersection of the social sciences (especially sociology, economics, political science, and organization theory) and information technology. His current projects include congressionally mandated studies on Internet navigation and the Domain Name System and improving cybersecurity research in the United States. His recently completed CSTB studies include *LC21: A Digital Strategy for the Library of Congress* (2001), *The Digital Dilemma: Intellectual Property in the Information Age* (2000), and *Trust in Cyberspace* (1999). Prior to joining CSTB, Inouye completed a Ph.D. from the School of Information Management and Systems at the University of California at Berkeley. In a previous life, Inouye worked in Silicon Valley, as a programmer (Atari Corporation), statistician and programmer/analyst (Verbatim Corporation), and manager of information systems (Amdahl Corporation). Inouye also completed other degrees—in information systems (M.S.), systems management (M.S.), business administration/finance (M.B.A.), liberal studies (B.S.), and mathematics (B.A.).

**MARJORY S. BLUMENTHAL** is the executive director of the Computer Science and Telecommunications Board—a 20-member Board of leaders from industry and academia—and its many expert project committees and staff. She designs, develops, directs, and oversees collaborative study projects, workshops, and symposia on technical, strategic, and policy issues in computing and telecommunications. These activities address trends in the relevant science and technology, their uses, and economic and social impacts, providing independent and authoritative analysis and/or a neutral meeting ground for senior people in government, industry, and academia. Marjory is the principal author and/or substantive editor of numerous reports and articles. The majority of her work has been cross-disciplinary. Before joining CSTB, Marjory was a manager of Competitive Analysis and Planning for GE Information Services. There she directed an analytical team supporting business development, product marketing, and field sales and developed business alliances for domestic and international network services. Previously, she was a project director at the former U.S. Congress Office of Technology Assessment, evaluating computer and communications technology trends and their social and economic impacts. There, among other things, she produced an internationally acclaimed study of computers in manufacturing and their implications for industries and employment. She is a member of the Santa Fe Institute Science Board, the Advisory Board of the Pew Internet & American Life Project, the TPRC Board of Directors, the editorial board of *ACM Transactions on Internet Technology*, and the ACM, AEA, and IEEE. In 1998 Marjory was a visiting scientist at the Massachusetts Institute of Technology, Laboratory for Computer Science. At MIT she developed and taught a course on public policy for computer science graduate students and pursued personal research interests. Marjory did her undergraduate work at Brown University and her graduate work (as an NSF Graduate Fellow) at Harvard University.

**MARGARET MARSH HUYNH**, senior project assistant, joined the Computer Science and Telecommunications Board in January 1999 and has worked on several projects. Currently, she is working on the projects on Internet navigation and the Domain Name System and the future of supercomputing. Ms. Huynh also assists with CSTB Board meetings and has worked on such recent projects as "Exploring Information Technology Issues for the Behavioral and Social Sciences," as well as those leading to the reports *IT Roadmap to a Geospatial Future* (2003), *Building a Workforce for the Information Economy* (2001), and *The Digital Dilemma: Intellectual Property in the Information Age* (2000). Ms. Huynh assists on other projects as needed. Prior to coming to the National Academies, Ms. Huynh worked as a meeting assistant at Management for Meetings for 4 months and as a meeting assistant at the American Society for Civil Engineers from September 1996 to April 1998. Ms. Huynh has a B.A. (1990) in liberal studies, with minors in sociology and psychology, from Salisbury State University, Salisbury, Maryland.

# B Briefers at Committee Meetings

## AUGUST 14-15, 2000
## NATIONAL RESEARCH COUNCIL
## WASHINGTON, D.C.

Steve Dietz, Walker Art Center
Jon Ippolito, Guggenheim Museum
Joan Shigekawa, Rockefeller Foundation
Bruce Sterling, Writer

## NOVEMBER 8-9, 2000
## AMERICAN INSTITUTE
## OF GRAPHIC ARTS
## NEW YORK CITY

Zoe Beloff, Artist
Kathy Brew, Thundergulch
Timothy Druckrey, Curator, Critic, and Writer
Robert Gehorsam, CBS/Viacom
Richard Grefé, American Institute of Graphic Arts
Mark Hansen, Bell Laboratories, Lucent Technologies
Perry Hoberman, Artist
Jaron Lanier, Advanced Network & Services Inc.
Daniel Lee, Bell Labs
Paul Miller (aka DJ Spooky), Music and Art Management Inc.
Warren Neidich, Artist
Daniel Oppenheim, IBM T.J. Watson Research Center
Anne Pasternak, Creative Time

Marah Rosenberg, Avaya Labs
Jakub Segen, Bell Laboratories, Lucent Technologies
Mark Tribe, Rhizome.org


# NOVEMBER 8, 2000
# NEW YORK UNIVERSITY
# CENTER FOR ADVANCED
# TECHNOLOGY
# NEW YORK CITY
# SITE VISIT

John Johnson, Eyebeam Atelier
Caroline Jones, Boston University
Neil Sieling, Media Arts Curator and Television Producer
Noah Wardrip-Fruin, New York University


# JANUARY 11-13, 2001
# STANFORD UNIVERSITY
# PALO ALTO, CALIFORNIA

Chris Chafe, Stanford University
Richard Gold,[1] Xerox PARC
Ken Goldberg, University of California at Berkeley
Kris Halvorsen, Hewlett Packard Laboratories
Barbara Hayes-Roth, Stanford University
Tim Lenoir, Stanford University and Extempo
George Lewis, University of California at San Diego
Michael Mateas, Georgia Institute of Technology
Robert Morris, IBM Almaden Research Center
Stuart Parkin, IBM Almaden Research Center
Arati Prabhakar, U.S. Venture Partners
Rick Prelinger, Prelinger Archives & Internet Moving Images Archive
Hal Varian, University of California at Berkeley
Gio Wiederhold, Stanford University

---

[1]Rich Gold (Richard Goldstein) of Menlo Park died in his sleep on January 9, 2003. Born on June 24, 1950, he received a B.A. from SUNY-Albany and an M.F.A. from Mills College.

# JANUARY 12, 2001
# PIXAR ANIMATION STUDIOS
# EMERYVILLE, CALIFORNIA
# SITE VISIT

Greg Brandeau, Pixar Animation Studios
Sharon Calahan, Pixar Animation Studios
Ed Catmull, Pixar Animation Studios
Rikki Cleland-Hura, Pixar Animation Studios
Tony DeRose, Pixar Animation Studios
Oren Jacob, Pixar Animation Studios
Randy Nelson, Pixar Animation Studios
Bill Reeves, Pixar Animation Studios
Tasha Wedeen, Pixar Animation Studios

# MAY 30-31, 2001
# MASSACHUSETTS INSTITUTE OF
# TECHNOLOGY
# MEDIA LABORATORY
# CAMBRIDGE, MASSACHUSETTS

Bruce Blumberg, Massachusetts Institute of Technology
John Guttag, Massachusetts Institute of Technology
Hiroshi Ishii, Massachusetts Institute of Technology
John Maeda, Massachusetts Institute of Technology
Rehmi Post, Massachusetts Institute of Technology

# JULY 29-30, 2002
# DANIEL LANGLOIS FOUNDATION
# THE CENTRE FOR RESEARCH AND
# DOCUMENTATION
# MONTREAL, QUEBEC, CANADA
# SITE VISIT

Alain Depocas, Daniel Langlois Foundation